Training & Reference Manual for Special Inspectors

Training & Reference Manual for Special Inspectors

Houman Parsaie, Ph.D.

Writers Club Press
San Jose   New York   Lincoln   Shanghai

Training & Reference Manual for Special Inspectors

Writers Club Press
an imprint of iUniverse, Inc.

For information address:
iUniverse, Inc.
5220 S. 16th St., Suite 200
Lincoln, NE 68512
www.iuniverse.com

ISBN: 0-595-20427-9

Printed in the United States of America

This book is dedicated to my lovely wife Evelyn and my Parents Forough and Iranpour.

CONTENTS

INTRODUCTION

This manual has been prepared for use as a reference materials for their day to day inspection business and for assistance in the training of new inspectors. This is also a supplement to applicable Standards, such as ASTM, ACI, AWS, etc. as well as building codes, such as UBC, SBC, etc.; thus, any references made in this manual reflects to the applicable code and/or standard test method.

Inspection is the observation of construction for conformance with the approved design documents. It shall not be relied upon by others as guarantee or acceptance of work, nor shall it in any manner relieve any contractor or other party from their obligations and responsibilities under the construction contract, or generally accepted industry custom, or building codes and standards.

Included in this manual are materials for other testing and inspection, for which there are currently no special training program or certifications available or offered.

H. John Parsaie, Ph.D.
Seattle, Washington

PREFACE

To prepare the trainees for use of materials in this manual or to provide the experienced special inspectors a quick and easy reference, following is Chapter 17, Section 1701, Structural Tests and Inspection, of the 1997 Edition of the Uniform Building Code (UBC):

Section 1701 – SPECIAL INSPECTIONS

1701.1 General: In addition to the inspections required by Section 108, the owner or the engineer or architect of record acting as the owner's agent shall employ one or more special inspectors who shall provide inspections during construction on the types of work listed under Section 1701.5

> EXCEPTION: The building official may waive the requirement for the employment of a special inspector if the construction is of a minor nature.

1701.2 Special Inspector: The special inspector shall be a qualified person who shall demonstrate competence, to the satisfaction of the building official, for inspection of the particular type of construction or operation requiring special inspection.

1701.3 Duties and Responsibilities of the Special Inspector: The special inspector shall observe the work assigned for conformance to the approved design drawings and specifications.
The special inspector shall furnish inspection reports to the building official, the engineer or architect of record, and other designated persons. All discrepancies shall be brought to the immediate attention of the contractor

for correction, then, if uncorrected, to the proper design authority and to the building official.

The special inspector shall submit a final signed report stating whether the work requiring special inspection was, to the best of the inspector's knowledge, in conformance to the approved plans and specifications and the applicable workmanship provisions of this code.

1701.4 Standards of Quality: The standards listed below labeled a "UBC Standard" are also listed in Chapter 35, Part II, and are part of this code. The other standards listed below are recognized standards.

1. Concrete:

    ASTM C 94, Ready-mixed Concrete

2. Connections:

    Specification for Structural Joints Using ASTM A 325 or A 490 Bolts-Load and Resistance Factor Design, Research Council of Structural Connections, Section 1701.5, Item 6

3. Spray-applied Fire-resistive Materials.

    UBC Standard 7-6, Thickness and Density Determination for Spray-applied Fire-resistive Materials.

1701.5 Types of Work: Except as provided in Section 1701.1, the types of work listed below shall be inspected by a special inspector.

1. **Concrete:** During the taking of test specimens and placing of reinforced concrete. See Item 12 for Shotcrete.

    EXCEPTIONS:

    1. Concrete for foundations conforming to minimum requirements of Table 18-I-C or for Group R, Division 3 or Group U,

Division 1 Occupancies, provided the building official finds that a special hazard does not exist.

2. For foundation concrete, other than cast-in-place drilled piles or caissons, where the structural design in based on $f'c$ no greater than 2,500 pounds per square inch (psi) (17.2 Mpa).

3. Nonstructural slabs on grade, including prestressed slabs on grade when effective prestress in concrete is less than 150 psi (1.03 Mpa).

4. Site work concrete fully supported on earth and concrete where no special hazard exists.

2. **Bolts installed in concrete:** Prior to and during the placement of concrete around bolts when stress increases permitted by Footnote 5 of Table 19-D or Section 1923 are utilized.

3. **Special moment-resisting concrete frame:** For moment frames resisting design seismic load in structures within Seismic Zones 3 and 4, the special inspector shall provide reports to the person responsible for the structural design and shall provide continuous inspection of the placement of the reinforcement and concrete.

4. **Reinforcing steel and prestressing steel tendons:**

    4.1 During all stressing and grouting of tendons in prestressed concrete.

    4.2 During placing of reinforcing steel and prestressing tendons for all concrete required to have special inspection by Item 1.

    EXCEPTION: The special inspector need not be present continuously during placing of reinforcing steel and prestressing tendons, provided the special inspector has inspected for conformance to the approved plans prior to the closing of forms or the delivery of concrete to the jobsite.

5. **Structural welding:**

    5.1 General: During the welding of any member or connection that is designed to resist loads and forces required by this code.

    EXCEPTIONS:

    1. Welding done in an approved fabricator's shop in accordance with Section 1701.7.
    2. The special inspector need not be continuously present during welding of the following items, provided the materials, qualifications of welding procedures and welders are verified prior to the start of work; periodic inspections are made of work in progress; and a visual inspection of all welds is made prior to completion or prior to shipment of shop welding:

        2.1 Single-pass fillet welds not exceeding 5/16 inch (7.9 mm) in size.

        2.2 Floor and roof deck welding

        2.3 Welded studs when used for structural diaphragm or composite systems.

        2.4 Welded sheet steel for cold-formed steel framing members such as studs and joints.

        2.5 Welding of stairs and railing systems.

    5.2 Special moment-resisting steel frames: During the welding of special moment-resisting steel frames. In addition to Item 5.1

requirements, nondestructive testing as required by Section 1703 of this code.

5.3 Welding of reinforcing steel: During the welding of reinforcing steel.

EXCEPTION: The special inspector need not be continuously present during the welding of ASTM A 706 reinforcing steel not longer than No. 5 bars used for embedments, provided the materials, qualifications of welding procedures and welders are verified prior to the start of work; periodic inspections are made of work in progress; and a visual inspection of all welds is made prior to completion or prior to shipment of shop welding.

6. **High-strength bolting:** The inspection of high-strength A 325 and A 490 bolts shall be in accordance with approved nationally recognized standards and the requirements of this section.

While the work is in progress, the special inspector shall determine that the requirements for bolts, nuts, washers, and paint; bolted parts; and installation and tightening in such standards are met.
(See UBC Chapter 17, Page 2-40 for exception to the above code.)

7. **Structural masonry:**

7.1 For masonry, other than fully grouted open-end hollow unit masonry, during preparation and taking of any required prisms or test specimens, placing of all masonry units, placement of reinforcement, inspection of grout space, immediately prior to closing of clean-outs, and during all grouting operations.

7.2 For fully grouted open-end hollow-unit masonry during preparation and taking of any required prisms or test specimens, at the start of laying units, after the placement of reinforcing steel, grout space prior to each grouting operation, and during all grouting operations. (See UBC Chapter 17, Page 2-40 for exception to the above code.)

8. **Reinforced gypsum concrete:** When cast-in-place Class B gypsum concrete is being mixed and placed.

9. **Insulating concrete fill:** During the application of insulating concrete fill when used as part of a structural system. (See UBC Chapter 17, Page 2-40 for exception to the above code.)

10. **Spray-applied fire-resistive materials:** As required by UBC Standard 7-6.

11. **Piling, drilled piers and caissons:** During driving and testing of piles and construction of cast-in-place drilled piles or caissons. See Items 1 & 4 for concrete & reinforcing steel inspection.

12. **Shotcrete:** During the taking of test specimens and placing of all shotcrete and as required by Section 1924.10 & 1924.11. (See UBC Chapter 17, Page 2-40 for exception to the above code.)

13. **Special grading, excavation and filling:** During earth-work excavations, grading and filling operations inspection to satisfy requirements of Chapter 18 and Appendix Chapter 33.

Section 1703 – Nondestructive Testing: Please refer to UBC Chapter 17, Page 2-41.

General Guidelines and Standard Test Methods

GENERAL GUIDELINES FOR FIELD INSPECTORS

A. General

1. Inspectors must protect themselves by strictly adherence to a strict moral and ethical code. Imprudent actions will follow you and eventually surface and harm your career.

2. Always present a professional image. Integrity must be faultless.

3. Discretion is a constant. Never accept favors or gratuities.

4. Avoid imprudent absences. If you are not there, you cannot verify work that has been completed. Many inspectors leave job sites while work is going on. This prohibited by laws.

5. Avoid ambiguity. Be specific in your requests to the contractor, engineer, and architect.

6. Provide adequate information. In writing your reports, be sure to pinpoint locations. In corresponding with the engineer, owner, or architect, put everything in writing and receive all change instruction in writing.

7. Cultivate satisfactory job relations and support open communications.

8. Execute decisions promptly. When you receive a change from the engineer, make sure everyone affected is notified immediately.

9. Do not control the work. You are not in charge of the work being done on the project. You must inform the person in charge of any violations in code, plans or specifications. You must also notify the engineer and governing agency. We do not buy concrete, steel or masonry materials. When a violation occurs, make adequate sam

ples and inform everyone concerned. Do not reject loads of concrete; let the contractor make that decision.

10. Maintain adequate records in your daily reports. Not all job problems, discussions, and changes. Recognize your responsibility in providing information for architect, engineer, etc.

B. Relations In The Field

1. Identify yourself.
2. Deal with the person in authority.
3. Arrive promptly for your appointment.
4. Phone if delayed.
5. Be courteous.
6. Be helpful.
7. Be businesslike.
8. Resolve differences without argument.
9. Refuse to accept favors.
10. Interpret the code the same for all, as you would under all similar circumstances.

C. Negative Inspector Qualities

1. Is argumentative has a short temper.
2. Uses his authority with a heavy hand.
3. Imposes personal prejudices rather than minimum standards.
4. Refuses to explain why.
5. Is vindictive and vengeful.
6. Is arbitrary and capricious.
7. Exceeds his limit of code enforcement responsibility:

8. Cost

9. Non-code matters

10. Personal involvement

D. Affirmative Inspector Qualities

1. Recognizes human limitations, including his own.

2. May be wrong and is not ashamed to admit it.

3. Is not too prideful to ask questions.

4. Tries to justify his decisions with defined reasons other than " that's what the code says."

5. Helps solve problems rather than condemning work, but does not attempt to change decision.

6. Discusses problems with other inspectors and comes to joint conclusions with all staff members.

8. Budgets his time.

9. Keeps appointments in a timely manner.

10. Keep adequate records.

11. Does not make demands he doesn't intend to enforce.

E. Tests and Inspections

Prior to the start of any project, the inspector should be prepared and aware of the required tests and inspections. In addition to tests required by project documents (Plans and Specifications), tests and inspections can be required the Building Department of the governing jurisdiction or the local building official. Special inspections required by building official typically follow inspections recommended in the Uniform Building Code and the Uniform Building Standards (STDS). Required special inspections are listed as a condition of the building permit and should be posted at the jobsite.

GENERAL GUIDELINES FOR EARTHWORK INSPECTION

OBJECTIVE

Earthwork as presented in this section includes, in general, those soils construction activities normally associated with grading excavation and filling. The purpose of earthwork observation and testing is to verify that the work is done in compliance with the approved plans and specifications and, in particular, the recommendations of the project geotechnical report.

Soil is a highly variable material, very sensitive to moisture fluctuations, and requires close attention to construction quality control in order to achieve the desired result. Many factors contribute to its suitability and effective performance. Identifying and properly controlling these factors can be divided into two general areas of activity. The first involves the observation or monitoring during construction with particular attention that placement and compaction operations are followed as specified in the contract documents and geotechnical report. The second involves tests to document the soils properties and verify compliance to quality specified.

Materials engineering laboratories that offer services in this field provide special expertise and equipment to verify the objectives of the design and project specifications. However, this is best accomplished when the Design Geotechnical Consultant provides these construction related services and can, in turn, achieve continuity and integration of the design-construct process. Without involvement of this Geotechnical Engineer, the constructed earthwork may not meet the performance requirements intended.

OBSERVATION DUTIES

A. Documents

1. Review plans, specifications, and the Geotechnical Engineer's report.
2. Note and record the equipment being used on site.

B. Sampling of Materials

1. Sample and verify that the following materials are delivered to the Materials Engineering Laboratory for any required testing:
   a) Subgrade materials
   b) Native-fill materials
   c) Imported materials, and
   d) Additive materials, (lime, cement, sand, pozzolan, etc.)

C. Testing

1. Perform soils classification and properties tests as required on native and/or imported soils.
2. Perform laboratory moisture-density relationship tests.
3. Where applicable, conduct a laboratory testing program to determine soils' properties resulting from admixtures such as cement or lime.
4. In the field, conduct in-place field density and moisture tests using procedures specified in the contract documents. Frequency of testing should be predetermined to allow for representative coverage of each lift, while interfering as little as possible with the earthwork operation's schedule.

5. Testing must be timely to avoid having to retest previously covered work. Similarly, test methods should be predetermined so as to take into account the Contractor's procedures and soil types.

6. Periodic sampling of materials in the field to verify continued compliance with specification requirements is recommended.

D. Reports

1. Submit written progress reports describing the tests and observations made and showing the action taken to correct non-conforming work.

GENERAL GUIDELINES FOR REINFORCING STEEL INSPECTION

A. Objective

The purpose of reinforcing steel observation is to give assurance that the supplier is exercising satisfactory control over production, fabrication and placing of reinforcing steel so that it meets the project specifications and applicable codes and industry standards.

This objective can best be achieved by qualified special inspectors who diligently perform the duties listed below while under the direct supervision of the material engineering laboratory.

B. Observation Duties

1. Documents
   a. Review the plans, specifications and approved shop drawings.
   b. Review applicable sections of referenced codes, such as: the Uniform Building Codes and Uniform Building Standards (UBC); the Building Code. Requirements for Reinforced Concrete (ACI-318) by the American Concrete Institute; the Manual of Standard Practice of the Concrete Reinforcing Steel Institute (CRSI); the Reinforcing steel Welding code (AWS D1.4) by the American Welding Society.

2. Mill Tests
   a. Verify reinforcing steel mill test reports (when available) for markings, test data, checking against project requirements.

b. Sample material for tests directly from unopened bundles when required by specifications.

3. Fabrication
   a. Check each shipment of reinforcing steel for the following
      1. Bar size and grades are as specified.
      2. Mill marking is in conformance with mill test report.
      3. Check for corrosion, contaminants, surface cracks and bars damaged in shipment.
      4. Check stop bends for specified radius and cracks.

4. Placement
   a. During placement of reinforcing, check for proper bar location, alignment, laps, ties, form and ground clearance, supports, field bends, radii, cracks, gouges or tack welds causing stress concentrations, removal of contaminants, and hardened concrete.
   b. If welding of reinforcing is required, it should be observed continuously, with particular emphasis on joint configuration, suitability of low hydrogen electrodes, preheat and interpass temperatures, and interpass slag removal. Check for welding and procedures for conformance to AWS D1.4.
   c. Prior to concrete placement, check for complete installation and notify contractor of any variations from plans and specifications. If variations are not corrected prior to start of concreting, immediately notify the design team representative for appropriate action.
   d. During concrete placement, check that reinforcing stays in-place and is adequately supported. Check for removal of soil, concrete spatter and grease.

5. Reports

   a. Submit written progress reports describing the tests and observations made and showing the action taken to correct non-conforming work. Itemize any changes authorized by architect/engineer. Report all uncorrected deviation from plans or specifications.

GENERAL GUIDELINES FOR CONCRETE INSPECTION

A. Objective

Concrete inspection as presented in this procedure refers to inspection and testing of fresh and hardened Portland Cement Concrete. The purpose of concrete inspection and testing is to verify that the work is done in compliance with the approved plans and specifications.
The objective of this procedure is to provide general guidelines for the concrete inspector to supplement the individual test procedures.

Because so many factors interact to affect the ultimate quality of concrete, it has earned a reputation as one of the most variable of construction materials. To deal properly with all these factors, quality assurance is divided into two easily recognized categories or phases. First involves collecting evidence from standard tests to demonstrate that the delivered concrete was produced to the quality specified. Second involves the enforcement of good construction practices during placement, finishing and curing to achieve a satisfactory finished product. These two objectives of quality assurance can best be achieved by qualified special inspectors who diligently exercise judgement in following the duties listed below while under the direct supervision of the materials engineering laboratory.

B. Observation Duties

1. Documents

    a. Review plans and specifications.

    b. Verify that the class of concrete ordered is being delivered and conforms with specifications, drawings and/or code requirements.

2. Observation Procedures

    a. Check forms for cleanliness and proper treatment prior to placement.

    b. Visually estimate the slump of each batch delivered and perform slump tests regularly.

    c. Determine concrete temperature, number of mixing revolutions and/or length of time batching.

    d. Observe placement procedures for evidence of segregation, possible cold joints, displacement of reinforcing or forms, and proper support of embedded items, anchor bolts, etc.

    e. Inspect for proper compaction/consolidation.

C. Sampling and Testing Duties

1. After a sample has been obtained the concrete shall be tested in the following order. After mixing the sample the temperature should be taken. The slump test should then be completed within 5 minutes after sampling. Then the unit weight and air are tested immediately following. All tests including concrete compression samples should be made within 15 minutes of obtaining the sample. *(Note that concrete used for air test should not be used in making samples).* Sample and test fresh concrete for the following (or as stipulated by plans and specifications):

    a) Slump.

    b) Entrained air.

    c) Temperature.

    d) Wet unit weight when required.

2. Sample concrete and prepare test cylinders in accordance with ASTM C31.

3. Field sampling and testing of concrete should be performed by a qualified technician, certified as an ACI – Grade I Concrete Field Testing Technician.

D. Reports

Submit written progress describing the tests and observations make and showing the action taken to correct non-conforming work. Itemize any changes authorized by architect/engineer. Report all uncorrected deviations from plans or specifications. Unless otherwise contracted for, concrete observation may not include verification of reinforcing, form dimensions or alignment, or finishing and curing procedures.

E. Hardened Concrete Sample Retrieval

Field made concrete samples are picked up on the day after the samples were made (24 hrs. {+\-8 hours} after molding). Samples molded on the day before a holiday or weekend are still picked up the following day, workday or not. Responsibility for assuring the concrete sample pick up is that of the testing and Inspection department supervisor. Compression test report forms, which are kept in the lab, also serve to identify samples which have not yet been picked up. Cylinders are transported in a padded box.

F. Concrete Sample Log-In

Concrete samples are stripped and logged in on the same day that they arrive in the lab. The sample ID number or cylinder code is a unique sequential number assigned to the sample set. Individual samples in the set are identified by the test date. Test date and special instructions are taken from the compression test report form filled out by the field technician. The samples are then placed in the cure tanks.

MAKING AND CURING CONCRETE TEST SPECIMENS IN THE FIELD

A In addition to the following procedure, refer to test method ASTM C31.

B Making Concrete Cylinders

When making concrete cylinders, three samples should be made. 1-7 day 2-28 days. This standard should be followed unless the job specifications state otherwise. Molds should be placed in an out-of-the-way place and on a solid surface such as a piece of plywood.

Rodding Technique (for slumps > 3 inches)

Place the concrete into the molds with a scoop in three equal lifts. Each lift should be rodded 25 times with the rod penetrating 1" into the lower lift. Between each lift, the sample is lightly tapped on the side with a hand to close any voids. After consolidation is completed strike off the surface of the cylinder with a trowel to form a smooth surface. The cylinder is tagged with the date cast, report # and Job #. It is also important to protect the cylinders from direct sunlight and hot and cold weather by storing in a cure box.

Samples with a slump of 3 to 1 inches

Either method may be used when the slump is in this rage.

Vibrating Technique (for slumps < 1 inches)

When making cylinders with a vibrator for consolidation it is important to keep an uniform vibrating time for that type of concrete, vibrator, and specimen mold. Also make sure not to over vibrate because this may lead to segregation. The concrete should be placed in the concrete mold in two equal lifts. Each lift should be vibrated with the sides of the cylinder lightly tapped after each vibration. The final lift should not exceed ¼" above the top of the cylinder. The surface shall be trawled off and then tagged with the following information; the date cast, report # and Job #.

The above procedures are the procedures for 12" cylinder molds. If a different size test cylinder is used check the chart below.

Cylinders Size	Mode of Compaction	# of Layers	Approximate Depth
12 inches	rodding	3 equal	4 inches
over 12 inches	rodding	as required	4 inches
12 to 18 inches	vibration	2 equal	half depth
over 18 inches	vibration	3 or more	8 inches

Diameters of Cylinders	Number of Strokes
6"	25
8"	50
10"	75

C. Making Concrete Beams

When making concrete beams, four samples should be made. 1-7 day 2-28 days and 1-hold. This standard should be followed unless the job specifications state otherwise. Molds should be placed in an out of the way place and on a solid surface such as a piece of plywood.

Rodding Technique (for slumps > 3 inches)

Place the sample into the beam in two lifts. For each lift make sure that the sample is rodded 60 times (1 rodding for each 2 square inches) and that the sides are sharply struck with a rubber mallet. The surface of the beam is then striked off and smoothed with a wooden float. The sample is then tagged with the date cast, sample number and the job number.

Samples with a slump of 3 to 1 inches

Either method may be used when the slump is in this range.

Vibrating Technique (for slump < 1 inch)

The beam mold is filled in one lift. The beam is then vibrated a uniform time for all beams made with that type of concrete vibrator, and specimen mold. Make sure not to over vibrate the concrete because that can lead to segregation of materials. After vibration the sides of the mold are sharply tapped with a rubber mallet. The surface of the beam is then striked off and smoothed with a wooden float. The sample is then tagged with the date cast, sample # and the job #.

The above information is for 6" beam molds. If using a different size mold please refer to the chart below.

Beam Size	Mode of Compaction	# of Layers	Approximate Depth
6 to 8 inches	rodding	3 equal	half depth
over 8 inches	rodding	3 or more	4 inches
6 to 8 inches	vibration	1	full depth
over 8 inches	vibration	2 or more	8 inches

After samples have been made make sure to keep surfaces wet with such items as dampened burlap bags, hay or blankets.

COMPRESSIVE STRENGTH OF CONCRETE CYLINDERS

A. In addition to the following procedure, refer to test method ASTM C39.

B. Procedure

Compression samples are removed from the cure tanks immediately before testing. Cylinders are broken as close to mid-day as possible in order to stay within permissible age tolerances (see table below).

Test Age	Permissible Tolerances
7-days	+/-6 hours
28-days	+/-20 hours

When testing, the samples are placed on retainer rings with neoprene pads on wiped clean plattens aligned with the top bearing block. The bearing block is rotated slowly by hand as the bottom bearing block is being raised. The bottom bearing block is raised up to the sample at fast-advance speed and fast-advance speed is maintained until the load on the sample has reached 10,000 lbs (approximately 350 psi) *Note that all samples are loaded to 20,000 lbs, at fast speed regardless of anticipated strength (except for samples with a maximum anticipated load of 40,000 lbs or less).* After reaching 20,000 lbs, the rate is lowered reduced to 35 psi/sec (1,000 lbs/sec for a 6" diameter sample). If a correction needs to be made in the load rate it is done immediately. Then the sample is loaded at a constant rate of strain until the sample yields.

After failure, the maximum load and the type of failure are recorded. For samples that break below the expected psi, observation of the fracture may also be noted.

C. Calculations

The compressive strength of each sample is calculated by dividing the maximum load by the cross-sectional area of the sample. Diameters are measured on at least 10% of the samples to be tested each day, with a minimum of 3 samples per day. The diameter of a sample is measured by use of a micrometer. This is used by taking a reading in the middle 1/3 of the cylinder. The reading is taken to the nearest 0.01". If the diameter average for the day does not fall with in the range of 6.00" +/- 0.02 then all of the cylinders for that day are measured. The cross-sectional area of all samples with an average of 6.00" +/- 0.02 is taken to be 28.27 in^2. All samples whose diameter falls outside this criteria have the individual diameter computed by the following formula:

$$Area = pi \ X \ (average \ diameter)^2 \ / \ 4$$

The length of the sample which appears to vary from the expected length of 12" by ¼" or more, shall be measured. Length to diameter correction (L/D) ratios shall betaken from the average length (average of four measurements) of the sample, as shown on the following chart.

L/D	12" Sample Length	Correction Factor
2.2+	13.200"+	Cut dons to 12"
1.8-2.2	10.800-13.200	1.00
1.69-1.8	10.125-10.800	0.98
1.56-1.69	9.375-10.125	0.97
1.46-1.56	8.750-9.375	0.96
1.37-1.46	8.250-8.750	0.95
1.29-1.37	7.750-8.250	0.94
1.23-1.29	7.375-7.750	0.93
1.19-1.23	7.125-7.375	0.92
1.15-1.19	6.875-7.125	0.91
1.10-1.15	6.625-6.875	0.90
1.06-1.10	6.375-60625	0.89
1.02-1.06	6.125-6.375	0.88
1.0091.02	6.000-6.125	0.87
<1.00	<6.00	Cannot be tested

Compressive strength shall be calculated by the following formula:

Compressive Strength = (maximum load/cross sectional area) x L/D factor

Lab Reports and Data Storage (Done by Lab Manager)

Each day's results are given to the person in charge of in-putting the compression test results. Under-strength results (or low breaks) are highlighted and brought to the attention of the project manager involved so that the deficiency may be brought to the attention of all of the involved parties immediately. Break results which are less the f'c at 28-days, or less then 70% f'c at 7 days are considered under-strength on most jobs. Daily break sheets include the following information.

A) Sample (or cylinder) code number

B) Project number of sample

C) Age of sample

D) Maximum load (lbs)

E) Compressive Strength calculated to the nearest 10 psi

F) Type of fracture (when other than cone)

G) Required psi at 28-days

H) Diameter of cylinder

I) Cross-sectional area of cylinder

J) Note any defects in cylinder or cap

The maximum load and compressive strength of all samples is logged into the permanent logbook that is retained in the lab.

UNIT WEIGHT, YIELD AND AIR CONTENT (GRAVIMETRIC) OF CONCRETE

A. In addition to the following, refer to test method ASTM C138.

B. Procedure

When this test is to be performed it should be done before the air content is determined in the air content measure. After weighing the empty measure, fill up the pot as for the air content test but before you continue with the air test, weigh the concrete and the measure. Then determine the weight of the concrete by subtracting the two numbers. The unit weight is then determined by multiplying this weight by the calibration factor for the measure. The yield is then determined by dividing the total weight of all material batched by the unit weight times 27. This number shows you the total yardage contained in the truck sampled. This is then compared by calculating the relative yield. This is the yield divided by the designed yardage. The closer the number is to 1 the more accurate the batch. The cement content (lb/yd3) can be determined by dividing the total pounds of cement by the actual yardage.

A = weight of measure
b=weight of measure and concrete
K=correction factor
c=unit weight
d=total weight of materials batched

e=yield of materials batched (yr3)
f=design yield of materials batched
g=relative yield of materials batched
h=weight of cement in the batch
I=cement content (lb. / yd3)

Unit weight ā = (b-a)Xk
Yield (yd3)(e)=d/(cx27)
Relative Yield (g)=e/f
Cement Content (lb./yd3)(I)=h/e

SLUMP OF HYDRAULIC CEMENT CONCRETE

A. In addition to the following, refer to test method ASTM C143.

B. Procedure

Slump is the first test. Dampen the mold and place it on the non absorbent plate. Next, step on the two foot peddles firmly. Begin to fill the slump cone 1/3 of the way by volume. (1/3 is a depth of 2 5/8 in: 2/3 is a depth of 5 1/4 inches). The sample is then rodded 25 times. The next 1/3 is added and rodded 25 times with the rod just penetrating the layer below. Fill the final 1/3 to the top and rod 25 times as before. After completing 3 lifts, strike off the top and clean any excess concrete from around the base. Make sure that the cone is held firmly against the plate at all times. Now remove the mold raising the cone 12 inches in 3-7 seconds. Then invert the cone next to the slumped sample and place the rodding bar across the top of the wide end of the mold. Measure the vertical distance from the bottom of the rod to the displaced center point on the slumped concrete. The slump should be measured to the nearest ¼ inch. If the sample shears off from one side of the sample, disregard that test and begin a new slump test. The slump test should be completed within 2 ½ minutes.

SAMPLING FRESHLY MIXED CONCRETE

A. In addition to the following procedure, refer to test method ASTM C172.

B. Procedure

The sample size that should be obtained is 1 ft^3 to make concrete cylinders. When running all of the tests it is recommended that approximately 1.5 ft^3 be sampled.

When sampling from stationary mixers, except paving mixers, a minimum of two samples should be taken at equal intervals. Obtain the sample by passing the sample container through the entire discharge stream or by diverting the entire discharge stream into the container. It is important not to restrict the flow of the concrete to prevent segregation. These samples are then mixed together to form one sample for the purpose of test.

When sampling concrete from a paving mixer obtain a least five different samples from the concrete that has been just discharged from the paver. Care should be taken to prevent contamination from the sub-base. This is done by placing containers in the areas to be sampled. The sample containers shall be shallow pans sufficient in size to obtain an appropriate amount of sample. These samples are then mixed together to form one sample for the purpose of testing.

Samples obtained from revolving drum mixers or agitators shall be sampled at two or more regular intervals. The samples should not be obtained from the very first or last portion of the batch discharge. This is sampled by passing the sample container through the entire discharge stream or by diverting the entire discharge stream into the container. Also samples should be taken after all water has been added. These samples are then mixed together to form one sample for the purpose of test.

This process should be completed in a maximum of 15 minutes.

MAKING AND CURING CONCRETE TEST SPECIMENS IN THE LABORATORY

A. In addition to the following, refer to test method ASTM C192.

B. Procedure for Lab curing

The cure tanks located in the laboratory are filled with tap water and are kept at a constant temperature in the range of (73.4 = / -3 deg. F) and are constantly recorded. Also the water level is checked daily and the "pH" is checked twice a week to make sure the tanks are saturated with lime. (pH = 10.7)

AIR CONTENT OF FRESHLY MIXED CONCRETE BY PRESSURE METHOD

A. In addition to the following procedure, refer to test method ASTM C231

B. Procedure

Place the measure on a level surface. Fill the measure in three equal lifts rodding each lift 25 times. When rodding the 2nd and 3rd lifts make sure the lower layer is penetrated approximately 1". Between each lift also make sure to smartly tap the sides 10 to 15 times with a rubber mallet. After consolidation is completed, strike off the top with the strike off bar.

After this is completed, clean the edges of the pot with your fingers. This is necessary for a tight seal between the lid and the pot. Wet the rubber gasket lightly and place the lid firmly on the pot. Make sure that all lid latches are tightly fastened to the pot. Next, make sure that the pressure valve is closed and open both petcocks. Close both petcocks. Next, pump the meter up to just above the initial pressure line. Let it sit for a few seconds and then open the pressure valve. Tap the sides of the pot sharply to relieve any local restraints. Lightly tap the side of the pressure gauge for the percent air content.

CAPPING CYLINDRICAL CONCRETE SPECIMENS

A. In addition to the following procedure, refer to test method ASTM C617.

B. Procedure

Capping is usually done the day before the samples are to be broken. Immediately prior to capping, the ends of the samples may need to be towel dried in order to allow good adhesion of the sulfur cap. Oil is used as a release agent between the sulfur cap and the capping mold. Oil is liberally applied to the capping mold and the excess is then wiped off with a cloth. After wiping off the excess oil, one ladle of sulfur capping compound is poured into the capping mold. The sample is then lowered onto the mold with a slight twisting/rotating motion as the sample comes into contact with the sulfur. This rotating motion helps push out any trapped air bubbles.

Samples with slightly deformed ends may be capped using standard procedures as long as the resulting cap is no more than 5/16" in thickness. If the resulting cap is more than 5/16" thick, the deformed end is to be cut square and recapped.

A minimum of 3 caps are verified for squareness and planeness after each day of capping. Squareness of caps is checked by use of a leveling plate and bubble level. Planeness of the caps is checked by the use of a straight edge and a 0.002" shim. The caps of 3 cylinders per day's capping are checked for planeness. Caps which are not square, plane, or contain voids are removed and replaced. Caps are checked for voids by tapping them with a screwdriver or hammer. Samples are returned to the cure tank after capping.

TEMPERATURE OF FRESHLY MIXED PORTLAND CEMENT CONCRETE

A. In addition to the following procedures, refer to test method ASTM D1064.

B. Procedure

When measuring the temperature of concrete make sure that the thermometer is a least 3" away from the sides of the sample container. Also make sure that the thermometer is at least 3" into the concrete. Keep the thermometer in the concrete until it reaches a constant temperature and record the temperature.

GENERAL GUIDELINES FOR CONCRETE BATCH PLANT INSPECTION

A. Objective

The purpose of batch plant observation is to verify that the concrete supplier is exercising adequate quality control to produce concrete that will meet the project requirements for materials, their batch proportions, mixing and adjustment for moisture.

This objective can best be achieved by qualified special inspectors who diligently perform the duties listed below while under the direct supervision of the materials engineering laboratory.

B. Observation Duties

1. Document
    Verify that the class of concrete ordered is being delivered and conforms with approved mix design.

2. Equipment
    a. Check the trucks for worn out or damaged fins, excessive build-up or hardened concrete, and for the presence of wash water from the previous delivery.
    b. Check the National Readymix Concrete Manufactures Association truck rating plate and verify the load capacities are not exceeded.
    c. Check the current "weights and measures" seal on scales.
    d. Verify that the moisture metering device is operable.

e. Verify that the scales start at and return to zero after each weighing operation.

f. Verify that the metering devices for admixtures have been calibrated recently and are operating.

C. Materials, Storage and Handling

1. Visually check the sand and coarse aggregate for method of storage, handling, source, grading, cleanliness, and moisture condition.

2. Obtain samples of aggregates when specified or when it appears that they may not conform to the required gradation or cleanliness.

3. Obtain grab samples of cement and pozzolanic materials when required by project specifications.

4. Check cement temperature.

5. For lightweight aggregates, check loose moist unit weight regularly and verify whether the plant is making proper adjustments to batch weights to compensate for variations in weight as well as moisture.

D. Batching of Materials

1. Record the volume in cubic yards for each class of concrete delivered. Verify that each mix proposed for delivery is of the proper designation and proportions approved for the project. Where discrepancies occur, request that the dispatcher clarify with the general contractor.

2. Verify that the specified materials are dispensed to the weight hopper and record the adjusted batch weights for all ingredients in the desired proportions of the concrete mix.

3. Verify that the proper adjustments have been made for variations in moisture of aggregates.

4. Record the mixing time and check whether it is sufficient.

5. Visually estimate the slump of the concrete and report immediately to the operator any measure outside specification.

6. Coordinate with the jobsite and verify the "as delivered" slump, air content, unit weight, mix temperature, general workability, and preparation of test samples.

E. Reports

1. Submit written progress reports describing the tests and observations made and show the action taken to correct non-conforming work. Itemize any changes authorized by architect/engineer. Report all uncorrected deviations from plans or specifications.

GENERAL GUIDELINES FOR PRE-TENSIONED CONCRETE INSPECTION

A. Objective

Because the strength of materials used in prestressed construction is significantly higher than normal concrete construction, a strong quality control program by plant manufacturers has developed. As a result, the purpose of pre-tensioned concrete plant observation is to verify the actual control program and check its effectiveness.

This objective can best be achieved by qualified special concrete inspectors performing the following duties under the direct supervision of the materials engineering laboratory.

B. Observation Duties

1. Documents
   a. Review the plans, specifications and approved shop detail drawings.
   b. Verify that concrete mix designs, tensioning data and calculations for stressing have been approved by the reviewing authority.
   c. Verify that jacking equipment has been calibrated.

2. Mill and Plant Test Reports
   a. Check performance of all materials to project specifications. Verify steel mill test reports for pre-stressing steel and deformed bar steel. Verify mill markings and tags. Verify cement mill test reports and certifications.

b. Check fabricator's testing facility and reporting of tests performed under fabricator's quality control program.

3. Sampling
   a. Sample and deliver or ship to the laboratory for testing the following when independent tests are required by project specifications:
      1. Concrete aggregates
      2. Pre-stressing strand or wire
      3. Reinforcing steel
      4. Steel used for structural steel embedded items
      5. Steel Fabrication of Embedded Items

4. Verify that qualified welders are employed to perform welding of structural steel using welding procedures qualified in accordance with AWS Structural Welding Code.

5. Pre-Placement Observation
   a. Bed layout and form cleanliness.
   b. Quantity and spacing of reinforcing and stressing steel.
   c. Location of inserts and embedded items.
   d. Profile of stressing steel.
   e. Witness tensioning of pre-stressing, elements, measure elongation of strand and record gauge pressure.

6. Test and Observation During Casting
   a. Perform batch plant observations.
   b. Conduct slump, air and unit weight tests. Request adjustments as necessary.

    c. Cast compression test specimens.

    d. Observe placement and vibration of concrete in forms.

    e. Observe finishing treatment.

7. Post-Placement Tests and Observation

    a. Observe curing procedures, temperatures and curing cycles.

    b. Monitor compressive strength results for specified release strength.

    c. Witness stress transfer.

    d. Identify member by component and date cast.

8. Field Erection

    a. Check members for damage during storage or shipment.

    b. Check field installation and structural connections.

9. Reports

    a. Submit written progress reports describing the tests and observations made and show the action taken to correct non-conforming work. Itemize any changes authorized by architect/engineer. Report all uncorrected deviations from plans or specifications.

GENERAL GUIDELINES FOR POST-TENSIONED CONCRETE INSPECTION

A. Objective

Post-tensioned concrete is normally constructed on-site rather than fabricated in plants. As a result, more responsibility is placed on the independent inspection agency to verify that quality control meets acceptable standards.

This objective can best be achieved by qualified special inspectors performing the following duties under the direct supervision of the materials engineering laboratory.

B. Observation Duties

1. Documents
   a. Review the plans, specifications and approved placing and stressing drawing furnished by the pre-stressing supplier.
   b. Review the reinforcing steel placing drawings to check whether they have been coordinated with the stressing drawings.

2. Mill Test Reports
   a. Check that reinforcing steel and pre-stressing steel supplied to job is properly identified and mill test reports show conformance to project specifications.

3. Sampling of materials
    a. Sample and deliver to the laboratory for testing the following materials when required by project specifications.
        1. Concrete aggregates and cement.
        2. Pre-stressing strand, rods, or wire.
        3. Reinforcing steel.
        4. Steel used for structural inserts.

4. Steel Fabrication of Embedded Items
    a. Visit fabrication plant.
    b. Verify that qualified welders are welding in accordance with AWS Structural Welding Code.
    c. Verify that only qualified welding procedures are being used.
    d. Observe the welding operations and the finished product for defects and verify that corrections are made, if necessary.

5. Pre-Placement Observation
    a. Check the general layout, size, spacing, and profile of all reinforcing steel and post-tensioning steel.
    b. Observe all anchorages, inserts, embedded items, blockouts, conduits, etc.
    c. Calibrate or review current calibration data on the proposed stressing equipment.

6. Observation During Placement of Concrete
    a. Observe batch plant operations when required.

b. Observe concrete placement and report any damage or misalignment of any embedded components (with particular emphasis at end anchorages).

c. Cast compression test specimens.

d. Test slump, air content and unit weight. Request adjustment as necessary

e. Check that mortar extrusions (fins) are cleaned off inside.

f. Check whether joints are tooled as specified.

g. Check required frequency of masonry wall prisms and observe construction of same.

h. Check for ties when specified.

i. Check horizontal reinforcing steel placing:

   a) Placed at correct course, laps as specified.

   b) Check whether laps are staggered in bond beams and corners as required.

   c) Check lintel bars over openings.

   d) Check hooks, if called for in jambs.

   f) Check ties in piers, diameter, spacing and properly fastened.

j. Check vertical reinforcing steel:

   a. Check bars a jambs, corners and piers, and typical wall steel.

k. Check whether tied at top and bottom and as required by project Grouting Observations

l. Verify that cells and starting beds are clean. Check condition with light or mirror.

m. Check whether dowels, anchor bolts and inserts are all in-place particularly at roof lines, floor lines and intersecting wall lines.

n. Check installation of clean-out closures.

o. Check grout mix and admixture required, etc.

p. Check slump in accordance with the specifications.

q. In low lift grouting, verify that maximum masonry height is in accordance with the code before grouting.

r. Check that grout is stopped below top for keying where required.

s. Verify mechanical vibrating during placement, and later during reconsolidation.

t. Continuous observation is required during grouting operations.

u. Prepare grout specimens in absorbent from, or as specified, for laboratory testing.

v. Check that curing requirements are being followed.

C. Reports

1. Submit written progress reports describing the tests and observations made and showing the action taken to correct non-conforming work. Itemize any changes authorized by architect/engineer. Report all uncorrected deviations from plans or specifications.

GENERAL GUIDELINES FOR MASONRY INSPECTION

A. Objective

The purpose of special observation for masonry is to verify that the workmanship and materials meet the minimum standards required by code as well as the project specifications. This is particularly difficult in masonry work where so much depends upon the capabilities of the individual mason as well as practices which have developed over the years and have become the custom of the trade for the particular locality. This requires experience and judgement by the inspector as well.

B. Observation Duties

1. Documents

    a. Review the plans and specifications with the masonry contractor and architect's representative in a preconstruction meeting to verify level of inspection required for the particular job. This is the time to resolve any differences in local custom or practice of the mason and requirements of the code and project specifications.

    b. Verify that mill test certification for unit masonry cement and reinforcing steel have been furnished by supplier and are acceptable to the architect/engineer.

41

C. Sampling of Materials

1. Sample and verify that the following materials are delivered to the laboratory for testing when required by project specifications:

    a) Concrete block or brick.

    b) Aggregates and cement for mortar and grout.

    c) Reinforcing steel as delivered.

D. Storage of Materials

1. Check that cement, lime, block and brick are supported on pallets and covered to protect from exposure to excessive moisture or drying.

2. Check that aggregates for mortar and grout are stored free from contamination and in such a manner as to minimize segregation.

E. Preparation for lay-up

1. Verify size and spacing of reinforcing dowels.

2. Verify that foundation concrete is clean and prepared as required by specifications.

3. Verify that lay-up or placing of masonry units have been approved for use.

4. Verify that cleanouts are provided for first course of each pour, if high lift method is used.

5. Check plumb and lay-up configuration.

6. Check moisture condition of masonry units.

7. Verify that proper mortar ingredients are batching techniques are being used and prepare mortar compression test specimens.

8. Check mortar time on board.

9. Verify that head joints are the same thickness as face shells or that full head joints are used when specified.

F. Reports

1. Submit written progress reports describing the test and observations made and showing the action taken to correct non-conforming work. Itemize any changes authorized by architect/engineer. Report all uncorrected deviations form plans or specifications.

G. For sampling and testing of Grout, please follow ASTM C1019 and UBC 21-18.

H. For field test specimens for Mortar, please follow UBC 21-16.

I. For compressive strength of masonry prisms, please follow ASTM E 447 and UBC 21-17.

GENERAL GUIDELINES FOR ASPHALTIC CONCRETE INSPECTION

A. Objective

The performance of asphaltic concrete pavement is as much affected by the careful construction of the subgrade and base as it is by the control of the asphaltic concrete itself. Therefore, the paving inspector must be knowledgeable in soils as well. The purpose of observation and testing of asphaltic concrete paving is to verify that the paving contractor and his supplier are exercising adequate quality control in their operations and are providing a finished project that complies with the project plans and specification requirements.

This objective can best be achieved by qualified special inspectors performing the following duties under the direct supervision of the materials engineering laboratory.

B. Observation Duties

1. Document and review approved plans and specifications and meet with contractor and suppliers before construction begins, to discuss project and to verify that requirements for testing and observation are well understood.

2. Review material certificates and test reports or compliance with job specifications.

3. Prepare or review mix designs for compliance to project requirements.

C. Sampling of Material

1. Sample and perform preliminary tests on proposed aggregates and asphalt cement (gradation, soundness, abrasion, stripping, etc.).

D. Subgrade and Base

1. Confirm that sources of materials have been sampled and approved.
2. Verify that materials delivered are of uniform quality.
3. Verify that control testing of subgrade materials is being performed and recorded as required.
4. Verify that subbase and base courses are of the source, type, thickness and density specified.
5. Verify that soil sterilization is provided if required.
6. Refer to Section 1, Earthwork, of this manual for additional details.

E. Batch Plant

1. The special inspector should become familiar with the appearance and physical characteristics of the mix to be used by observing visually the finished mixture so that unsatisfactory conditions may be readily recognized.
2. Check the batch plant facilities prior to production of asphaltic concrete mixture.
3. Check aggregates in stockpile to verify conformance to materials utilized in the design.
4. Check the bin weights of the aggregate fractions and asphaltic cement (batch plant only).
5. Check the temperature of the mixed batches on the truck.

6. Perform hot-bin gradations of the blended aggregates (where applicable).

7. Verify cold-bin feeds and hot-bin batch weights are adjusted as necessary to produce the job-mix formula within tolerance.

8. Before loading, truck beds should be checked for cleanliness and absence of materials that might be detrimental to the mix.

9. Coordinate with the jobsite inspector to obtain a uniform and consistent asphaltic concrete mixture.

F. Spreading and Paving

1. The field inspector should contact the batch-plant inspector promptly should conditions be observed during placement and spreading operations that suggest a need for change at the plant. The following items should be addressed prior to and during placement operations:

   a) Area to be paved, cleaned and properly primed, or tack coated.

   b) Leveling course installed where required.

   c) Suitability of spreading and paving equipment.

   d) Asphalt mix temperature when delivered and after final rolling, is within what is required.

   e) Density tests by nuclear gage during rolling.

   f) Thickness control by adequate placement and compaction.

   g) Sampling of asphaltic concrete at jobsite during placement for laboratory testing extraction, gradation, stability, etc.).

   h) Core samples taken for verification of thickness and density of in-place asphaltic concrete.

i) Application of seal coat and curing in accordance with specification requirements, if required.

G. Verification Tests

1. Stability and density, bulk specific gravity and maximum specific gravity.
2. Asphalt content by extraction.
3. Aggregate gradation of the mixture from extracted sample.
4. Physical properties of the asphalt cement: penetration, viscosity, ductility, and plasticity index, and sieve analysis.
5. Aggregate quality: Los Angeles abrasion, liquid limits of soils and plastic limit and plasticity index , and sieve analysis.
6. Field density.
7. Thickness determination.
8. Smoothness tolerance.

H. Reports

1. Submit written progress reports describing the tests make and showing the action taken to correct nonconforming work. Itemize any changes authorized by architect/engineer. Report all uncorrected deviations from plans or specifications.

GENERAL GUIDELINES FOR SHOTCRETE INSPECTION

A. Objective

The purpose of special observation for shotcrete is to verify that the materials, processes and the particularly unique application techniques conform to the project documents. The process moves rapidly in often noisy and congested environments: it relies heavily on experienced working crews.

The quality control objectives can best be achieved by a thoroughly experienced special inspector who understand shotcrete as an extension of his or her concrete inspection knowledge and is under the direct supervision of a qualified materials engineering laboratory.

B. Observation Duties

1. Documents
   a. Review plans, specifications and contractor submittal for applications process used.
   b. Verify crew qualifications.
   c. Verify test methods and sample procedure.
   d. Verify test methods and sample procedure.

2. Observation Procedures
   a. Verify main and auxiliary equipment for compliance, capacity, pressures and proper functioning.
   b. Check for hot oil cold weather limitations and precautions.

c. Verify reinforcing has been previously inspected and placed for minimal congestion.

d. Verify all joints, penetrations, embeds and formwork are correct and adequately supported.

e. Verify the nozzleman has suitable shooting positions and access to achieve placement with minimal rebound.

f. Check for ground wires or other thickness gauging control method.

g. Review mixing and placing procedures with crew before commencement of application.

h. Observe placement for:

1. Consistency

2. Consolidation

3. Coverage

4. Rebound

5. Finish

C. Sampling and Testing

1. Prepare a test panel 18" x 18" x 3", or as otherwise specified to obtain suitable cores for testing. Arrange correct positioning of sample panel to represent job shotcrete. Prearrange with nozzleman the correct timing of the test sample preparation and verify that it is representative of job placement, finish and cure. Refer to ACI 506 for further guidance.

2. Mark panel with specimen identification and protect for curing period.

D. Reports

1. Submit written progress reports describing the tests and observations made and showing the action taken to correct non-conforming work. Itemize any changes authorized by architect/engineer. Report all uncorrected deviations from plans or specifications.

GENERAL GUIDELINES FOR STRUCTUR-AL STEEL INSPECTION

A. Objective

The customary practice of fabrication of steel in the shop prior to erection conveniently allows division of observation of structural steel into two basic categories, shop and field. While the purpose is to assure that proper quality control is exercised at each location, the environment differs. Often the shop is fabricating other projects concurrently and may operate two or three shifts per day. The shop work is closely related to mass production, while the field work related closer to handcrafting.

These duties should be performed by qualified special inspectors under the direct supervision of a materials engineering laboratory. To better achieve the objective of quality assurance, it is wise to use only one agency to fulfill the duties of both shop and field observation.

B. Observation Duties

1. Review plans, specifications and approved shop drawings
2. Review applicable sections of referenced codes, particularly the American Welding Society Structural Welding Code (AWS D1.1).
3. Review welding procedure qualifications when other than standard AWS pre-qualified joints and procedures are involved.

C. Mill Test Report

1. Review mill test reports and check heat numbers with material as received. Verify that proper identification of steel is maintained during fabrication.

D. Sampling and Testing

1. When required by project specifications (particularly schools and hospitals), mark sample location with steel stamp on each piece tested.

2. Record sample number and location and check that sample identification is maintained as samples are delivered to laboratory and tested.

3. When steel members are delivered to finish length and no "crop ends" are available for sample cutting, coordinate cutting and patching requirements with architect/engineer and fabricator.

E. Welding Observation (Applicable to Shop and Field)

1. Check each welder's certification and verify that the welder does work only as covered by his certification.

2. Keep a written record of each welder by name, his identifying steel mark and the percentage of rejectable welds.

3. Upon detection of a rejectable weld (either visually or by nondestructive test) the inspector in charge will notify the welder and/or his foreman for verification of defect. The inspector –in-charge will observe removal of defects and repairs to check whether acceptable procedures were used.

4. Check structural members for thickness adjacent to welds.

5. Inspect joints for proper preparation, including bevel, root faces, root opening, etc.

6. Check the type and size of electrodes to be used for the various joints and positions. Check the storage facilities to see if they are adequate to keep the electrodes dry.

7. Observe the technique of each welder periodically with the use of a welding inspection shield.

8. Verify the use of proper per-heat and interpass temperatures.

9. Observe multi-pass welds continuously. Continuous observation is defined as follows: The inspector is present in the welding area at all times. The extent of inspection of individual welds will depend on the number operators welding.

10. Observe single pass fillet welds periodically, after determining that the operator is capable of producing the welds required.

11. If straightening or restraining of weldments is necessary, verify that approved methods will be used.

12. Tag or stamp accepted weldments with the inspector's identification stamp.

F. Workmanship (Shop)

1. Check straightening and bending procedures.

2. Check cut edges, including those frame cuts, shears or milled.

3. Check bolt holes in major connections for size.

G. Additional Field Duties

1. Discuss welding sequence for general construction plans and specific joint sequence with steel contractor and engineer to verify proper sequence to minimize restraint.

2. During adverse weather conditions, check that adequate steps are taken to prevent moisture penetration at welding locations.

H. High-Strength Bolting

1. Sample high-strength bolts, washers, and nut for testing from the lots in the shop or on the jobsite.
2. Check calibration of wrenches for tightening capacity in a wrench calibrator.
3. Check joints surfaces to verify that they are free of burrs, dirt, etc.
4. Review the procedure for installation of bolts. The amount of inspection during installation will depend on the method used.

I. Painting

1. Verify cleaning operations for all surfaces are to condition specified.
2. Verify conformance of paint to specifications.
3. Verify conformance to application method, brush, roller or spray.
4. Check for thickness of each coating, and final thickness.
5. Check touch-up for final finish conformance.

J. Reports

1. Submit written progress reports describing the tests and observations made and showing the action taken to correct non-conforming work. Itemize any changes authorized by architect/engineer. Report all uncorrected deviations from plans or specifications.

GENERAL GUIDELINES FOR NONDESTRUCTIVE TESTING

A. Objective

The purpose of nondestructive testing is to verify that structural steel and/or completed welds are sound with respect to the given project criteria. Visual observation may not detect hidden fusion defects, cracking and laminar tearing. Therefore, it is important that all means necessary be available to the special inspector for reasonable verification of sound welds.

This objective can best be achieved by qualified NDT special inspectors performing standard test methods under the direction of the materials engineer laboratory. Since NDT tests are indirect (relying on a probing medium to disclose defects), accurate evaluation depends upon experienced, qualified personnel who are thoroughly trained and applications.

B. Observation Duties

1. Review plans, specifications and approved shop drawings.
2. Review applicable section of referenced codes and standards, particularly UBC Section 2722(j) and Section 6 of the AWS Structural Welding Code D1.1.
3. Where applicable, review welding procedures and sequences.

C. Personnel

1. All NDT personnel shall be qualified in accordance with the American Society for Nondestructive Testing, Recommended Practice

SNT-TC-1A and the supplement applicable to the method being used. Only Level II and III inspectors or Level I inspectors working under the direct supervision of a Level II or III inspector are permitted to conduct the test.

D. Method Selection

1. Method to be used shall be as prescribed by project specifications, building codes, or as recommended by the materials engineering laboratory under the direction of design professional.

2. Effective use of NDT depends on utilizing the proper test method and techniques. Where field conditions or sequences affect the specified methods, the NDT technician will make recommendations for suitable approved methods or techniques.

E. Tests

1. Perform tests as prescribed by contract documents, for welds, lamination or laminar tearing.

2. Upon detection of a defect, mark the defect, and notify the foreman and/or the lead visual inspector.

3. Keep written records of pieces, welds, welder identification marks, length and location of defects, method and date of repair, number of retests, records of performance of each welder (percent of rejected welds), sampling rate, etc.

F. Reports

1. Submit written progress reports describing the tests and observations made, their locations, and any corrective actions taken.

2. Report the current percent of rejectable welds.

G. Standards

1. Many nondestructive testing standards and codes are presently available for information and reference. Most standards and codes specify equipment and personnel requirements, operational steps and acceptance standards tied to the end-use function. Following is a partial list of the more common standard test methods.

   a) Radiography – AWS D1.1, ASTM E94 and E99, ASME Section V.

   b) Ultrasonic Testing – AWS D1.1, ASTM E164, ASME Section V.

   c) Magnetic Particle Testing – ASTM E109, ASME Section V.

   d) Penetrant Testing – ASTM E165, ASME Section V.

GENERAL GUIDELINES FOR SPRAY-APPLIED FIREPROOFING INSPECTION, PART I

A. Objective

The purpose of spray-applied fireproofing observation is to verify that the application of material is in accordance with the project specifications, applicable codes, and manufacturer's recommendations.

This objective can best be achieved by experienced special inspectors who diligently perform the duties listed below while under the direct supervision of the materials engineering laboratory.

B. Observation Duties

1. Documents

   a. Review plans, specifications and manufacturer's recommendations.

   b. Review applicable section of referenced codes and standards, such as the Uniform Building Code and Uniform Building Standards (ICBO).

2. Observation Procedures

   a. Verify substrate condition for cleanliness prior to application.

   b. Verify application in accordance with codes and specifications.

C. Testing and Sampling Duties

1. Measure thickness of spray-applied fireproofing in accordance with specifications and Uniform Building Standard 43-8.

2. Remove and deliver samples to materials engineering laboratory for unit weight tests.

3. Re-inspect areas repaired due to insufficient thickness or damage by sampling, tenant improvements, panel placement, rain, etc.

D. Reports

1. Submit written progress reports describing the tests and observations made and showing the action taken to correct non-conforming or damaged work. Itemize any charges authorized by architect/engineer. Report all uncorrected deviations from plans and specifications.

GENERAL GUIDELINES FOR SPRAY-APPLIED FIREPROOFING INSPECTION, PART II

References:

American Society for Testing and Materials [ASTM]: ASTM E 605 & E 736

Uniform Building Code 1701, 703, 704, UBC Standard 7-6
Seattle Building Code: SBC Chapter 43, DCLU Director's Rule 15-90

Definition:

Sprayed Fire-Resistive Materials [SFRM]: Fiber and cementitious type materials that are sprayed onto substrates to provide fire-resistive protection of the substrate.

Approved Drawings:

SFRM shop drawings must be prepared by the contractor, submitted to and approved by the Engineer of Record prior to application. All inspections shall be made in accordance with the approved shop drawings which clearly identify the SFRM material and the thickness of SFRM material for the *primary, secondary, and other structural frame members.*

Structural members that provide lateral bracing for the structure frame but carry vertical loads are included in the definition of primary structure when they are essential to the stability of the building as a whole.

Inspected areas shall be marked on a reduced plan provided by the owner and maintained on the job site for inspector's use as well as reference.

Prior Special Inspection:

All substrates such as welds, bolts, etc. must be inspected and approved by the special inspector prior to application of SFRM.

Special Inspection Procedure:

- The SFRM materials shall be verified at site as conforming to the type and brand indicated *and* approved on the SFRM shop drawings.

- The condition of the substrate shall be inspected prior to application of the SFRM. The substrate shall be prepared according to the manufacturer's instructions and shall be free of dirt, grease, oil, loose paint and primer, and others that prevent adequate adhesion.

- For thickness measurement, use a depth gauge consisting of a movable needle or pin and a disc perpendicular to the needle. The thickness shall be measured by inserting the penetrating pin touches the substrate, the disc shall be moved against SFRM with sufficient force to register the average plane of the surface. Then, withdraw the gauge to read thickness to an accuracy of 1/16".

- Where the specified thickness is in excess of ½ inch, a minus tolerance no greater than 1/8 inch shall be permitted. Where the specified thickness is ½ inch or less, no minus tolerance is permitted.

- A thickness to density correction formula is contained in certain fire resisting rating criteria or is available from some SFRM manufacturers. Make sure to consult the rating criteria and the SFRM manufacturer before citing for deficiency.

- Measure SFRM thickness and compute the average of test results as outlined in the following table. Also, see *ASTM E 605* and *UBC Standard 7-6* for detailed sketches of locations to be measured for each type of structural member.

Type of Member	Frequency, each Floor	Test Locations
Beams	1. 25% of total number of members 2. The greater of 25% of the members requiring each thickness or three [3] of each thickness.	Nine locations at each end of a selected 12 inch length
Columns	1) 25% of total number of members 2) The greater of 25% of the members requiring each thickness or three [3] of each thickness.	12 locations at each end of a selected 12inch length
Decks	10 measurements for each thickness for each 10,000 sq.ft. randomly.	Five locations for flat deck; four locations for fluted deck [valley, crest, two sides]; all in a 12 inch square area
Trusses [Joists]	1) 25% of total number of members 2) The greater of 25% of the members requiring each thickness or three [3] of each thickness.	7 locations at each end of a selected 12 inch length
Non-Structural beams, columns, trusses, etc.	1) 10% of total number of members 2) The greater of 10% of the members requiring each thickness or three [3] of each thickness.	7 locations at each end of a selected 12 inch length
Pipes, conduits, etc.	Mechanical, electric, and plumbing installations shall not be embedded in SFRM	N/A

- The random areas selected for test measurements shall be marked on the reduced floor plan prior to inspection. Results of test measurements are to be recorded on data sheets. Test locations on columns and beams are to be selected at the end thirds or middle thirds, in a rotating order to vary the locations of test areas.

- In general, visually check all structural frame members and floor sections on each floor. Areas appearing to be less than required thickness, damaged areas, and dropouts are to be checked for thickness and marked for recoating, where required.

- All patching of damaged areas including density test locations shall be inspected no more than 24 hours prior to being covered.

- Final inspection of exterior surface shall be performed until cladding is completed. SFRM shall not, upon complete drying or curing, show any deep or wide cracks, voids, spalls, or any exposure of the substrate.

- Conduct one density test at random on each of the following protected elements, per floor or per 10,000sq.ft., whichever provides the greatest number of tests: the flat portion of the deck [1]; a beam [2], either the bottom of the beam lower flange or the beam web; and a column [3], either the column web or the outside of one of the column flanges.

- Use a rectangular template of a known area and sample the test area. At least five [5] measurements shall be taken within sampled area. One measurement shall be taken at the center of the test area and one at the four corners approximately 1-1/2 inches from adjacent sides. Thickness measurements shall be performed prior to removing the thickness of the specimen.

- The Specimen shall be cut along the perimeter of the template. All of the in-place material shall be carefully removed from the substrate and delivered to the laboratory for lab density test.

- No sample shall have a density less than 5% below the specified density. If so, it needs to be corrected to the satisfaction of the building official. *UBC Standard 7-6* and/or *ASTM E 605 8.2.1.1* acceptance criteria may be used to get approved by the building official.

Working Tools:

1. Steel Rule, graduated in at least 1/16 inch intervals.
2. Thickness gauge.
3. Knife or other suitable device for cutting the specimen.

GENERAL GUIDELINES FOR GLU LAM AND TRUSS JOISTS INSPECTION

A. Objective

The fabrication of most glu-lam and truss joist products if conducted in controlled plant conditions which are designed for a mass-produced product. The primary purpose of observing the product at the plant is to check the critical operations such as gluing and to provide verification that the quality control exercised by the fabricator is adequate.

To best achieve this objective an experienced timber technician should be employed performing the following duties under the direct control of the engineering materials laboratory.

B. Observation Duties for Glu Lam Timber

1. Documents
   a. Review plans, specifications and approved shop drawings.
   b. Review applicable sections of referenced codes, particularly the Timber Construction Manual by the American Institute of the Timber Construction (AITC) and reference standards of the Uniform Building Code by ICBO.
   c. Verify that the proposed lumber grades, combinations, adhesive, and end joint details meet the code requirements.

2. Materials
   a. Verify certification on lumber grading, adhesives and preservatives.
   b. Verify lumber grade marks on the pieces being used.

3. Observation Requirements—Preliminary
   a. Verify that shop drawings have been reviewed and stamped by Architect/Engineer and General Contractor.
   b. Verify that spacing of joints meets job and code requirements.
   c. Measure moisture content of lumber and verify with acceptance range specified.
   d. Check appearance and grade requirements.

4. Observation of Sub-Assemblies (End Joints)
   a. Verify lumber grade at end joints.
   b. Gluing and curing procedure, verification of following:
      1. Lumber moisture, temperature and cross-section.
      2. Workroom humidity and temperature.
      3. Adhesive certification, lot and temperature.
      4. Joint match and separation.
      5. Assembly temperature, pressure and time.
      6. Sample and test representative joints.

51

      7. Laminating (Gluing)
   a. Recheck lumber grades, combinations and faces, moisture and temperature.
   b. Record workroom temperature and humidity.

    c. Adhesive certification, lot verification and temperature.

    d. Verify camber assembly.

    e. Gluing and curing:

        1. Observe glue spread and check for skips.

        2. Record open time prior to clamping.

        3. Record clamping pressure.

        4. Record curing temperature and time.

        5. Sample and test (block shear, core shear, cyclic de-lamination).

5. Finishing

    a. Recheck joint spacing and cross-sectional dimensions.

    b. Observe repairs for appearance.

    c. Record and inspect surface treatment.

        1. Preservative

        2. Sealer

        3. Primer and paint

    d. Hammer-break each member, prepare shipping certificate.

    e. Observe and record wrapping.

6. Reports

    a. Submit written progress reports describing the tests and observations made and showing the action taken to correct non-conforming work. Itemize any changes authorized by architect/engineer. Report all uncorrected deviations from plans or specifications.

C. Observation duties for truss-type joist construction

1. Chord Fabrication

   a. Perform all requirements of "Glu Lam Timber Observation Duties".

   b. Check end joint spacing at panel points.

   c. Check drilling and routing for webs

2. Web Fabrication

   a. Structural Steel:

      1. Review specification requirements.

      2. Review mill certification, steel and coating.

      3. Sample and test steel when specified.

3. Fabrication

   a. Verify web wall thicknesses and diameters at specified locations.

   b. Check for splitting at flattened ends.

   c. Check alignment edge distance and pin-placement.

   d. Check bridging clips, bearing clips and ridge connectors.

   e. Check truss dimensions

   f. Check connector welding if performed.

4. Reports

   Submit written progress reports describing the tests and observations made and showing the action taken to correct non-conforming work. Itemize any changes authorized by architect/engineer. Report all uncorrected deviations from plans or specifications.

GUIDELINES FOR SPECIAL INSPECTION OF ANCHOR BOLTS, DOWELS, AND HOLDOWN SYSTEMS INSTALLATIONS

This is a materials guideline for field inspectors engaged in the inspection of anchor bolts, dowels, and holdown system inspections, since there is no special inspection program for such inspectors.

✿ ASTM and other Code resources, references and materials

ASTM Standards:

ASTM A 325:	Specification for Bolts, Steel, Heat-Treated, 120/105 ksi minimum Tensile Strength (15.08)
ASTM A 490:	Specification for Heat-Treated Steel Structural Bolts, 150 ksi minimum Tensile strength (15.08)
ASTM F 680:	Test Method for Nails (15.08)
ASTM F 1575:	Test Method for Determining Bending Yield Moment of Nails (15.08)
ASTM F 1554:	Specification for Anchor Bolts, Steel, 36, 55, and 105 ksi Yield Strength (15.08)
ASTM E 488:	Test Methods for Strength of Anchors in Concrete and Masonry Elements (04.07)
ASTM E 754:	Test Method for Pullout Resistance of Ties and Anchors Embedded in Masonry Mortar Joints (04.07)

ASTM E 1512:   Test Method for Testing Bond Performance of Adhesive Bonded Anchors (04.07)

1997 Uniform Building Code

97 UBC Vol. II, Chapter 17

97 UBC Vol. II, Chapter 18
97 UBC Vol. II, Chapter 23, Divisions I, II, III, IV, and V
97UBC Vol. II, Chapter 25
2000 International Building Code

2000 IBC, Chapters 17, 18, 23, and 25

Note: For Single family residences less than 5000 sq. ft. total living area, please refer to the newest publication of International Residential Code (IRC). For any single dwelling with 5000 sq. ft. or more, refer to either 97 UBC or 2000 IBC.

✿ The Most Common Used Anchorage Systems

a) **HILTI:**   HIT/HAS   ICBO-ES # ER-5193 (6/95)

HVA   ICBO-ES # ER-5369

b) **ITW:**   RAMSET/REDHEAD EPCON SYSTEM (mostly used on the

east coast)   ICBO-ES # ER-4285

c) **COVERT:**   Injection Adhesive Anchors and Undercut Anchors
ICBO-ES # ER-4846

d) **Earth Bound:** Seismic Holdown Systems Using the IMPASSE DEVICE
ICBO-ES # ER-5378

e) **SIMPSON ANCHORING SYSTEMS**: (most commonly used throughout
United States)   ICBO-ES #: various numbers

✿ Testing

A minimum of 5% of resisting tension anchors, threaded rods, and through-bolts shall be tested in accordance with the procedures described in ASTM E-488-90, with a minimum of two (2) tests required.
Where wall thickness varies, a minimum of one test shall be performed on an anchor which has the minimum embedment. Tests must show that bolts can maintain a tensile load of 3000 pounds (13.35 kn) for at least five (5) minutes with an allowable deviation of no more than 10%.

A minimum of 25% of installed anchors resisting tension *and* shear shall be tested utilizing a calibrated torque wrench. The torque for the 1/2 inch (12.7 mm) diameter anchors, the 5/8 inch (15.9 mm) diameter anchors, and the 3/4 inch (19.1) diameter anchors is 40 foot-pounds, 50 ft-lbs., 60 ft-lbs. (54.2 N-m, 67.8 N-m, and 81.3 N-m), respectively.

No visible deflection or deformation is allowed during the above-noted torque testing process.

A minimum of 25% of installed threaded rods, and reinforcing bar anchors resisting shear shall be tested utilizing a calibrated torque wrench. The torque for the 3/4 inch diameter (19.1 mm) rod and no. 6reinforcing bar is 60 foot-pounds (81.3 N-m), for the no. 5 & 4 reinforcing bars, the torque is 50 ft-lbs. And 40 ft-lbs. (67.8 N-m, and 54.2 N-m), respectively.

✿ Report Writing

All reports shall contain the following required minimum data:

1) Date of Inspection

2) Permit Number

3) Project Name

4) Project Address

5) Name of the Special Inspector

6) Test Location

7) Hole Depth, Diameter, and Cleanliness

8) Product Description (product name, rod, diameter and length, adhesive expiration date, etc.)

9) Torque Test Results (minimum of 25% shall be tested, see TESTING), to include the applied load to each anchor/rod

10) Pullout Test (tensile load test) Results (minimum of 5% shall be tested, see TESTING), to include the applied load to each member

11) Verification of Anchors Installation with the manufacturer's published instructions and procedures, which is usually included in each package, and if applicable, refer to the Evaluation Reports, tested and published by the International Conference of Building Officials (ICBO-Evaluation Service) appropriate report number, as well as applicable ASTM standards, including but not limited to ASTM E 488-90.

STANDARD TEST METHOD FOR PENETRATION RESISTANCE OF HARDENED CONCRETE

This is an Introduction to one of the nondestructive examination devices for hardened concrete strength testing, which is approved and governed by the above mentioned ASTM test method, WINDSOR PROBE. This introduction is designed to help Inspectors who have previous experience performing this type of test as well as Inspectors without the necessary experience. However, it is *STRONGLY* recommended that the less experienced Inspectors perform this type of testing at least once or twice with an experienced Inspector so they can have quality hands-on experience as well as the opportunity to get their specific technical questions correctly answered. This way most of the questions and/or concerns can be properly addressed.

This material will cover the basic procedures for performing this test and operating WINDSOR PROBE.
In order to get familiar with this standard test method (or Code), please refer to the *ASTM C 803-90, C 670* and *ANSI A10.3*. This material is based on the above mentioned standards as well as the manufacturing hand book for this test method *(DENSICON Handbook for WINDSOR PROBE model #532CF-A)*.

1. Referenced Documents:

a) ASTM Standards: C 803-90 & C 670

b) ANSI Standard: A10.3 *Safety requirements for Powder Actuated Fasting Systems*

2. Preface and history about this type of tests:

In the concrete construction industry, it is vitally essential to the contractor, Engineer, Architect and Owner to know for a certainty that structural strength requirements have been achieved not only during but also after the construction cycle.

The safety of construction workers and ultimately the structure's occupants and traffic are involved. Equally important is the integrity and continuity of the principles in their industry.

Concrete is usually purchased on the basis of 28 day laboratory cured cylinders. This practice has been adequate and effective in the past when relatively low strengths were specified and there was enough attention paid to concrete testing at the job site. However, in recent years with the introduction of improvements in cement, chemical admixtures and pumped concrete, there has been a marked increase in the use of high strength concrete and corresponding decrease in the attention paid to on site testing. As a result, the number of failures have increased and the number of requests each year for strength verification of *in-place* strengths is also increasing.

There are other methods of testing in-place concrete for strength. One non-destructive method is the use of the Impact Hammer [also known as SCHMIDT concrete test HAMMER]. This instrument, when in proper calibration and in the hands of a skilled technician, is a valuable tool in evaluating concrete strengths. The Impact Hammer does have some deficiencies, however, and is also subject to manipulation by an unskilled and/or unscrupulous user.

The only other quantitative non-destructive on site test method available, recognized and approved to the concrete industry is a ballistic type test system, which is called the *WINDSOR PROBE TEST SYSTEM.*

The Windsor Probe Test System is a method of measuring penetration resistance of steel probes driven into plain or reinforced concrete by an accurately controlled powder charge. The calculation of compressive strength is based upon a previously established relationship between the type of coarse aggregate in the concrete to be tested and the exposed length of the driven probes.

This system is the only fast reliable and accurate test method available to us today to readily established actual compressive strength in-place. The probes can be used in the horizontal or vertical position with little surface preparation of the structure to be tested.

3. Preliminary Statistical Analysis:

Statistics predicate that one cylinder, core, probe or any single incident is a measure only of itself. If two measurements are made, and are different, it is difficult or impossible to determine which is correct. For this reason, a Precision Statement sets forth a standard deviation, and when used with three results, establishes the likelihood of the true average which may be the two highest or the two lowest of a set of three.

4. The Windsor Probe Test System:

There are two types of Probes; a silver colored for standard weight concrete [over 125 lbs./cu/ft] and a gold colored Probe for lightweight concrete [less than 125 lbs./cu/ft], and only one precision power load that may be used at two power levels, standard power or low power. In other words, Silver colored: step diameter, used for natural stone coarse aggregates and Gold colored: Straight diameter, used for lightweight aggregate, i.e. expanded shale.

When the concrete is less than 28 days, always use the Low Power setting as described in Table No. 2. If the probes are loose and not firmly embedded or results exceed 4000 psi, use Standard Power setting and read from he attached Table 1.

When using the probe at Standard Power and results indicate less than 3200 psi, disregard and use Low Power.

Unless a simple indication of curing progress is required, always set **three [3] Probes** and use the average of the three as result unless the extreme variation exceeds 0.200 inches. If the difference exceeds this measurement, set a fourth probe and discard the one that deviates most from the average.

In all cases, test the coarse aggregate of the mix to determine the correct MOHS' column to convert Probe height to psi [Mohs' tables are included in this manual]. Don't be concerned that the aggregate may vary. If so, Precision Statement adherence will detect bad readings. Also, if you know as to who delivered the concrete in question, you can call the concrete supplier and ask them about their coarse aggregate [Mohs' number].

5. Observe these Precautions:

a) Never attempt to operate the driver for any other purpose than probe testing concrete.

b) Do not load the driver until you are ready to test.

c) Do not attempt to use the driver on any material other than concrete or materials with sufficient density to resist excessive penetration of the probe.

d) Never attempt to operate the driver except in combination with the locator template.

e) Do not point the driver at yourself or others.

f) Do not let bystanders gather around. Bystanders need to stay behind you.

g) Use goggles when operating the driver.

h) Do not use the driver with an obstruction in the barrel.

i) Do not use the driver where inflammable vapors may be present.

j) Always use ordinary common sense precautions consistent with all high energy devices.

6. Limitations of the Windsor Probe:

Probes at Standard Power cannot be used with accuracy unless these conditions are followed:

a) Not closer than four [4] inches from an unsupported edge at Standard Power level.

b) Not closer than 2 ½ inches from an unsupported edge at Low Power level.

c) Never on 6 X 12 inch cylinders or 6" x 8" x 28" flexure beams when at Standard Power level. Test only 8" x 8" x 28" beams, or use Low Power for 6" x ^'.

d) Do not use on thin slabs if the Probe will penetrate more than half the thickness of the slab when using proper power level.

e) If surface is coarser than a broom finish, grind smooth.

f) If the Probe must be used for concrete containing maximum size coarse aggregate approximately 2 inches, a number of Probes may be loose and rejected by the Precision Statement limits [for Precision Statement please refer to the paragraph 7 in this manual].Excessive variation will also exist when coring.

g) A certificate of calibration is furnished to the user when the equipment is delivered. At least once a year, the system components should be recertified for calibration. This may be done at job site or the equipment can be returned to the factory. All recalibration in-plant is within 24 hours of receipt.

h) All Windsor Probe System operators should be certified by a factory trained Instructor and receive an "**ACREDITED OPERATOR**" or "**CERTIFIED OPERATOR**" card. This is an one day course with an open book exam at the end of the course.

7. Precision Statement for ASTM C 803:

a) These numbers represent indexes of precision as described in ASTM C 670 for Preparing Precision Statements for Test Methods for Construction Materials. The Precision of probe measurements has been found to vary with the maximum size of coarse aggregate. The indexes of precision given in the following table apply to measurements on the same concrete, made with the same materials , procedures equipment, and curing conditions and obtained by a single operator using the same instrument.

Maximum Size of Aggregate	[1S] b)	Max. range of 3 individual measurements c)	[D2S] d)	Max. Diff. Between 2 averages of 3 measurements e)
No. 4 [Mortar]	0.08 in.	0.26 in.	0.23 in.	0.13 in.
1-in.	0.10 in.	0.33 in.	0.28 in.	0.16 in.
2-in.	014 in.	0.46 in.	0.40 in.	0.22 in.

b) These numbers are the single-operator standard deviations for concrete made with the maximum size aggregate shown in column 1.

c) These numbers are the maximum allowable ranges for groups of three individual measurements made closer together, either as individual measurements or using the triangular

positioning device. If the range of three measurements exceeds the limit given, the three should not be averaged. In this case, a fourth probe should be fired, and the one that deviates most from the average of the four should be discarded. If the three remaining measurements still do not meet the given limit, the device should be moved to a different area and three new measurements obtained.

d) These numbers are the maximum difference between two individual measurements [D2S limits]. These indexes should not be used if measurements are being obtained in groups of three and averaged.

e) A difference larger than the amount given indicates a probability that there is a significant difference in the concrete in the two areas represented by the two groups of three measurements each.

8. Helpful hints about Probing:

- Probe testing hardened concrete provides results of curing rate and in-place strength developed by the placement conditions peculiar to the concrete involved. It is a test of the concrete in its permanent structural environment at the time of test.

- Do not expect the in-place strength to be identical to the strength indicated by lab or field made cylinders, particularly at early ages less than 28 days. Field cylinders made at the job site cure at a different rate due to their small mass as compared to large placements. Lab cylinders are also protected from the elements, which effect job concrete. However, at ages more than 28 days, if good placement controls have been achieved, 28-day lab cylinders results will correlate very well with in-place probe results.

- Do not expect job site cores to correlate with in-place strength unless they were taken in accordance with ASTM C 42 and ACI 318-95, which states that cores should be broken *dry* if the structure will not be more than superficially wet. *Wet* cores will break 15-20% lower than *dry* cores [ACI 318-95, Chapter 504]. Another 15% may be added according to ACI 318-95, Section 4.3.5.1.

- Concrete strength results vary from front, middle and end of the truck depending upon the time between loading and discharge. To expect in-place strength results to be consistent throughout the structure is unrealistic. A variation of 10-20% coefficient of test variation is normal for well placed concrete with good control practices [ACI 214, Standards of Concrete Control, Table 2].

- The Windsor Probe conversion Tables 1 and 2 were developed independently of each other. For new or low strength concrete [less than 3200 psi], the Low Power technique is used. The conversion table 2 was calibrated for the specific probe velocity at Low Power; Table 1 was calibrated to the specific velocity of the probe at Standard Power. The latter is used for mature, high strength concrete in excess of 3200 psi [usually over 28 days old].

- The Low Power Table 2 was calibrated independently from Standard Power Table 1. There is no direct relationship. If Standard Power is used and reflects strength lower than 3200 psi, the results will reflect values lower than actual indicating that Low Power should be used. The same difference in results will be obvious when press breaking identical cylinders if the press velocity is substantially changed.

- When probing standard 6" x 12" cylinders, always use Low Power and place cylinders in a Windsor Vee Calibration Fixture available for this purpose. Cylinders over 3500 psi cannot be reliably probed due to insufficient mass to resist the Probe penetration at the high energy delivered by Standard Power.

• If Probes do not penetrate concrete, check to be certain the Driver Head has been screwed firmly on to the Probe prior to inserting in the barrel of the Driver.

• In new concrete placed during hot summer months, do not be concerned if in-place concrete develops 3000-3500 psi as early as 24 hours. This is the result of "sun-air" curing that would not effect the small cylinders. Frequently, the moisture in a field cylinder will evaporate rather than contribute to the hydration and curing of the cement in the cylinder. Consequently, a 24 hour field cylinder will be much lower than the in-place strength [ACI 301-95 Field Reference Manual and NRMCA Publication number 170, ASTM Standards for Concrete Technician Certification].

• Under winter concreting conditions, the in-place concrete may cure at a much slower rate than a field cylinder held in a curing box. Also, please note that in winter months, when the temperature falls below 32 degree F., chances are that the concrete in-place is frozen. In this condition, do not use the Windsor Probe System due to the danger of injury by performing the test.

• Cores drilled from 40-50 year old concrete will almost always break lower than the actual in-place concrete strength. It has been established that the minute cracks at the interface of the aggregate/paste have a tendency to propagate due to drilling vibration. This, in combination with the core being removed from the restraint of the reinforcing steel and the load of the structure will cause a misleading low strength indication. Conversely, the Probe may not completely detect a weakened bond at the aggregate interface. The correct strength of such old structures is likely to be somewhere between core strength and Probe strength and the data should be interpreted by the Project Engineers and/or Department Manager.

- *Round Gravel:* When stream rounded gravel is used for coarse aggregate, the bond at the paste aggregate interface is not always firmly developed at early ages and results in low press breaks which do not compare well with the actual aggregate to study and identify this condition. Correlation to in-place strength is even more difficult when this situation exists.

- *High Strength Concrete:* Probe Strength Tables were calibrated from in-place test data. Lab controlled tests and other correlation to conventional test specimens such as drilled cores or cylinders, properly reported according to ACI, are usually excellent up to 4000 psi strength levels. However, it is increasingly difficult to consistently make accurate 6000 psi or higher cylinders or cores due to the many inherent variables [60 reasons that effect cylinder strength results, ACI Journal]-Table 1. As strength increases, the difficulties in making accurate cylinders or drilled cores increases. At these high strength levels, the Probe is unaffected and the cylinders or cores should be carefully analyzed if lower than the in-place strengths are indicated.

- *Loose Probes:* When Probing concrete in excess of 6500 psi, certain types of concrete will not retain the probe. If this condition exists, clean the hole made by the Probe with a hand bulb syringe and measure the depth of the hole, then subtract from 3.125 inches [Probe length]. Use this dimension to convert to psi [table 1].

- *Firmness of Probe Embedment:* Tap all Probes for verification of final seating and correction of any minor rebound.

9. Mohs' Hardness Scale:

A universally accepted system for identifying minerals in terms of hardness. All minerals on earth are classed in ten groups: Numbered from 1 to 10. Number 10 [hardest] is the DIMOND.

Number 1 [softest] is TALC. Each higher number stone [mineral] will scratch the number lower than itself.

10. Procedure to Identify Concrete Aggregate for Use of the Windsor Probe Test:

a) Select a piece of aggregate that is exposed in the area of test.

b) Scratch the aggregate to be identified with the #9 stone of the Mohs' table.

c) If the scratch mark will not wipe off the aggregate, use #8, #7 etc. until the mark does wipe off. [Example: If a #6 scratch can be wiped off, the aggregate is a #6]

11. Where to Probe:

These are in respect to ASTM C 803 as well as manufacturing recommendations

- *Rectangular Suspended Slabs:* First typical floor, place three Probes in center of slab topside to establish a norm, plus a single Probe at each corner. The center three may be eliminated on subsequent floors. It is recommended that one set of probes [3] for each 5000sq. ft. of stripping area.

- *Architectural Shaped Slabs:* Place at least one Probe test in each structural element or each 3000 square feet.

- *Columns or Walls:* If a three Probe test is used singly, place one at least shoulder height from floor; one higher and one lower with a minimum spacing of 7 inches.

- *Probing through Forms [Columns or Walls]:* When necessary, Probes may be driven through wood forms or steel [up to 1/16 inch thick] with no allowance for strength loss. Add the form thickness to the measured height or Probe. In this case, do not use the single probe measuring base plate or probe cap.

- *Probing Core-Type Continuous Plank for Cutting Strength:* Calibrate to cylinders cured under identical curing conditions. Cap cylinders immediately after initial set, strip, Probe three of six and press break the remaining three while still in the moist condition. Do not let cylinders remain uncovered in open air. Then use single Probes, one at each end and one at center.

- *Hardened Surfaces:* Probe normally. Surface treatments that are less than 3/16 inches thickness do not effect the Probing. The results will indicate the strength of the base concrete, not the surface plate. When thicker coatings exist, probe the underside of the slab if accessible.

- *Lightweight Concrete:* Be sure to use gold color Probes and read results from Mohs' Column No. 3, Table 1 if Standard Power is used; Table 2 if Low Power is used.

- *Footers or Pier Caps:* Probe topside [bearing surface]. Be sure the surface is at least broom finish or grind.

- *Post-tensioning Cables:* Use single Probes in center of element and one at each corner to insure same or equal batching or placement. If any doubt exists, Probe within two feet of the bulk head.

- *Architectural or Tilt-up Slabs:* When cast horizontally, Probe singly at two corners and center diagonally. The topside is the compression side when "picking" vertically.

- *Pre-stressed Beams:* Place single Probes topside over web at each end or fifty foot intervals to insure that bond to cables has been achieved prior to detensioning.

- *Fire Damaged Concrete:* Set at least three [3] sets of three [3] Probes in an unaffected area and use the combined average as a norm. Then place in sets of three [3] in a grid pattern of 1000 square foot areas. If results in affected area exceed 0.160 inches in Probe height average, the concrete has been affected, either by increase or decrease in strength.

- *Frozen Concrete:* Always start with Low Power. Use appropriate Probe depending upon concrete unit weight. It is, however, strongly recommended to avoid testing frozen concrete.

- *Concrete Block Testing:* When testing block made for use in structural masonry walls or pre-stressed floor plank, calibrate the Probe by single probing ten [10] blocks. Use Low Power, Probe over the web and press break the same ten. Thereafter, Probe one block at the beginning and end of every four [4] hours production or at the end of any continuous run if less than four hours.

- *Concrete Pipe:* Develop a norm of three [3] Probes for each Class pipe. Thereafter, use one Probe in center, outside if tamper packer-head made and inside if centrifically spun. Usually, once or twice during a days production is adequate unless equipment conditions change.

- *Probing Standard Cylinders:* Always use Low Power to probe cylinders. Place Probe in center of the 12 inches height. A Vee Calibration Fixture is available for this purpose. Please note that cylinders in excess of approximately 3500 psi cannot be reliably probed. The mass is insufficient to resist the Probe energy if used at Standard Power. Cracking and/or spalling will result.

- *Probing Standard Flexure Beams:* Standard 6" x 6" x 28" beams can be probed singly if the strength is not over 3500 psi and will receive Probes at Low Power. Always probe topside of beam, preferably before removal from form. Only 8" x 8" x 28" beams can be probed at Standard Power.

12. Advantages of the Windsor Probe

- *Contractors:* Testing, as such, is not too important to a Constructor; his lab provides this. The Probe is a method to shorten stripping and form removal cycles and keep job schedules. For example, when a low cylinder break is reported, this might cause a series of extremely expensive delays. Ownership and use of a Probe may prevent this. The Constructor, in fact, owns the building until a Certificate of Occupancy is issued. When delays occur, the constructor can determine, for his own protection, whether to delay or proceed.

  A typical situation is that perhaps only one-fourth of a slab is low due to "tired" concrete left too long in the last truck of the placement. This need not cause a delay in stripping if the contractor can quickly locate the low strength area. He will simply simultaneously reshore the questionable section.

  Low strength due to slow curing is not a problem to a constructor if he can delineate the area. The remedial methods are well known to successful contractors.

- *Concrete Precast Producers:* COST SAVINGS. The Precaster is usually faced with the problem of getting increased production from molds or forms. Profits are in direct relationship to production. Because of the inherent problems in making test cylinders to insure adequate stripping strength, the mix is nearly always overdesigned. The Probe provides the ability to test the product and permits monitoring of the curing rate to establish stripping strength. As a result, less expensive design mixes can be developed over a very short period of time and substantially reduce cost.

- *Concrete Pipes:* The major problem of the Pipe producer is not ultimate strength; it is early strength to achieve rapid production rates in removal from the molding equipment and early transportation. With the use of steam curing techniques, this is attainable in the product but extremely difficult to develop in a test cylinder due to the difference in compaction. Certain procedures, such as vibrator tables, tend to remedy this but the product compaction and cur ing conditions are not accurately duplicated. It is not uncommon to see 800 to 900 pounds of cement used when only 600 pounds of cement would suffice if the product itself is tested.

- *Prestress Producers:* The major problem in this industry is to achieve production from molds. This can only be achieved by insuring that detensioning strength specifications of the bond to stressed cables has been developed. This elusive property is most difficult to isolate due to steam or electric curing of the product because test cylinders do not cure at the same rate.

  It is not uncommon to see the cylinder wrapped in electric blankets or "cookers" to simulate the product conditions. These difficulties result in overdesign "just to be safe". The Probe will test such products accurately and subsequent monitoring will permit less costly design mixes.

- *Commercial Testing Labs:* The Probe is **NOT** intended to replace design-mix 28 day cylinders. It is an additional profitable service to offer the clients. When you receive a panic call from your client due to a low cylinder break, it is not in his best interest to suggest coring out of habit. His problem is immediate and your use of the Probe in his behalf will solidify your Lab/Customer relationship. [I must confess that the test results of coring concrete samples and testing the samples for compression strength are more accurate and closer to the actual strength of the questionable concrete.]

  The Commercial Testing Labs historically are similar to the legal profession. Testing Labs make their available services known but intend to shy away from advertising or soliciting. Consequently, the old techniques are suggested out of habit and/or lack of knowledge rather than reason, and both the Lab and the Client incur financial loss. A **MODERN** Lab should provide probing services, ultrasonic pulse velocity, Pachometer, Floor flatness/levelness devices and all of the newer techniques available.

- *Engineering Firms:* The major supportive influence for the Windsor Probe and other in-place testing methods stems from the Engineering and Architectural Professions. They represent the owner and their integrity is "on the line" even after completion of a structure.

  Many Engineering firms own Probe testing equipment to confirm job test reports prior to signing off for the Certificate of Occupancy.

  Another major use is to evaluate existing structures for insurance companies and to develop data for upgrading or designing additions to older structures.

- *Ready-Mix Concrete Producers:* The standard cylinder, made at the point of delivery, is the only record this industry has. It is, in truth, a test of concrete at the change of ownership. The producer knows from experience that it is only a matter of time until a low break report causes serious doubt as to the quality of his product. Usually, it is simply a case of a faulty cylinder, not the delivered product. Consequently, the producer is compelled to overdesign and incur excessive cost simply to insure against the fallibilities inherent in standard testing.

- *City Agencies:* In most larger cities, every construction agency, such as the Transit Authority, Department of Transportation and Department of Sewer and Water Resources, Port Authority, Department of Education, etc. own and use Windsor Probes to confirm test data.

STANDARD POWER-TABLE NO. 1

➤ **Important Instructions:**

This Table is used only for the STANDARD POWER range of the Windsor Probe System, operated in accordance with the manufacturers Instruction Manual. The Table represents the results of calibrating the system to the velocity of the probe at the STANDARD POWER position. STANDARD POWER is used for testing concrete, in existing structures, usually cured longer than 28 days. ALWAYS change to LOW POWER if the Probe System, used at standard power, indicates less than 3000 psi. This Table, No. 1, has no fixed relationship to Table No. 2. Each Table has been calibrated independently to the respective probe velocity. A point of convergence will occur in the range of 3600 psi, and may vary slightly, depending on the design mix. Please note that if the speed [velocity] of a crushing press was changed for breaking standard cylinders, a separate calibration formula for computing psi would also be required. Always confirm the Mohs' Number of the coarse aggregate with a Mineral Scratch test [as described in this manual] or calibrate the system to standard cylinders. For standard weight concrete [>125 lbs./cu. ft.], use silver color RRS-01 [1/4 inch diameter probe] and read results in appropriate Mohs' column from Table No.1. For lightweight concrete [<125 lbs./cu. ft.] use gold color PRS-03 [5/16 inch diameter probe] and read results in Mohs' Number 3 column from Table No.1. For mortar [no coarse aggregate concrete], use silver color RRS-01 and read results in Moh's Number 3 column from Table No.1.

EXPOSED PROBE [inches]	COMPRESSIVE STRENGTH [p.s.i.]				
	Mohs' No. 3	Mohs' No. 4	Mohs' No. 5	Mohs' No. 6	Mohs' No. 7
1.275	-	-	-	-	-
1.300	-	-	-	-	-
1.325	-	-	-	-	-
1.350	-	-	-	-	-
1.375	-	-	-	-	-
1.400	3000	-	-	-	-
1.425	3175	-	-	-	-
1.450	3325	-	-	-	-
1.475	3500	-	-	-	-
1.500	3675	3000	-	-	-
1.525	3825	3175	-	-	-
1.550	4000	3350	-	-	-
1.575	4175	3525	-	-	-
1.600	4325	3700	3050	-	-
1.625	4500	3875	3225	-	-
1.650	4675	4050	3400	-	-
1.675	4825	4225	3600	-	-
1.700	5000	4400	3775	3000	-
1.725	5175	4575	3950	3200	-
1.750	5325	4750	4150	3400	-
1.775	5500	4925	4325	3600	-
1.800	5675	5100	4500	3800	3000
1.825	5825	5275	4700	4000	3225
1.850	6000	5450	4875	4200	3425
1.875	6175	5625	5050	4400	3650
1.900	6325	5800	5250	4600	3875
1.925	6500	5975	5425	4800	4100
1.950	6675	6150	5600	5000	4300
1.975	6825	6325	5800	5200	4525
2.000	7000	6400	5975	5400	4750
2.025	7175	6675	6150	5600	4975
2.050	7325	6850	6350	5800	5175
2.075	7500	7025	6525	6000	5400

2.100	7675	7200	6700	6200	5625
2.125	7825	7375	6900	6400	5850
2.150	8000	7550	7075	6600	6050
2.175	8175	7725	7250	6800	6275
2.200	8325	7900	7450	7000	6500
2.225	8500	8075	7625	7200	6725
2.250	8675	8250	7800	7400	6925
2.275	8825	8425	7975	7600	7150
2.300	9000	8600	8175	7800	7375
2.325	9175	8775	8350	8000	7600
2.350	9325	8950	8525	8200	7800
2.375	9500	9125	9725	8400	8025
2.400	9675	9300	8900	8600	8250
2.425	9825	9475	9075	8800	8475
2.450	10000	9650	9275	9000	8675
2.475	-	9825	9450	9200	8900
2.500	-	10000	9625	9400	9125
2.525	-	-	9825	9600	9350
2.550	-	-	10000	9800	9550
2.575	-	-	-	10000	9775
2.600	-	-	-	-	10000

STANDARD POWER-TABLE NO. 2

➢ **Important Instructions:**

This Table is used only for the LOW POWER range of the Windsor Probe System, i.e. the probe is positioned 2-1/2 inches downstream in the driver barrel. The Table represents the results of calibrating the system to the low velocity of the probe at the LOW POWER position. ALWAYS use the low power range for testing concrete less than 28 days after placement, or until the concrete has cured sufficiently to cause loose probes [approx. 3800 to 4500 psi]. If the probes are not firmly embedded change to Standard Power. This Table, No. 2, has no fixed relationship to Table No.

1. Each Table has been calibrated independently to the respective probe velocity. Always confirm the Mohs' Number of the coarse aggregate with a Mineral Scratch test [as described in this Manual] or calibrate the System to standard cylinders. For standard weight concrete [>125 lbs./cu. ft.], use silver color RRS-01 [1/4 inch diameter probe] and read results in appropriate Mohs' column from Table No. 2. For lightweight concrete [130 to 120 lbs./cu. ft.] use gold color PRS-03 [5/16 inch diameter probe] and read results in Mohs' Number 3 column from Table No. 2 or apply the appropriate correction factor shown in the Table below:

Lbs./cu. ft.	Correction Factor
130 to 121	100% of Mohs' No. 3 column
120 to 115	84% of Mohs' No. 3 column
114 or less	64% of Mohs' No. 3 column

For mortar [no coarse aggregate concrete], use silver color RRS-01 and read results in Moh's Number 3 column from Table No.1.

EXPOSED PROBE [inches]	COMPRESSIVE STRENGTH [p.s.i.]				
	Mohs' No. 3	Mohs' No. 4	Mohs' No. 5	Mohs' No. 6	Mohs' No. 7
1.125	525				
1.150	625				
1.175	725	450			
1.200	800	525			
1.225	900	600			
1.250	1000	675			
1.275	1075	750			
1.300	1175	825	450		
1.325	1250	900	525		
1.350	1325	975	600		
1.375	1400	1075	700		
1.400	1500	1150	775		
1.425	1575	1225	875	400	
1.450	1650	1300	975	500	
1.475	1725	1400	1050	600	

1.500	1850	1500	1150	700	
1.525	1925	1575	1250	800	
1.550	2000	1675	1325	900	450
1.575	2075	1750	1425	1000	550
1.600	2150	1850	1525	1100	650
1.625	2250	1950	1600	1200	750
1.650	2325	2025	1700	1300	875
1.675	2400	2100	1800	1400	975
1.700	2500	2200	1875	1500	1075
1.725	2575	2275	1975	1600	1175
1.750	2650	2375	2075	1700	1275
1.775	2750	2450	2150	1800	1400
1.800	2825	2550	2250	1900	1500
1.825	2900	2650	2350	2000	1600
1.850	3000	2725	2425	2100	1725
1.875	3075	2800	2525	2200	1825
1.900	3150	2900	2625	2300	1925
1.925	3350	2975	2700	2400	2050
1.950	3325	3075	2800	2500	2150
1.975	3400	3150	2900	2600	2250
2.000	3475	3250	2975	2700	2375
2.025	3550	3350	3075	2800	2475
2.050	3650	3425	3175	2900	2575
2.075	3750	3500	3250	3000	2700
2.100	3850	3600	3350	3100	2800
2.125	3925	3675	3450	3200	2900
2.150	4000	3775	3525	3300	3025
2.175	4075	3850	3625	3400	3125
2.200	4150	3950	3725	3500	3250
2.225	4250	4050	3800	3600	3350
2.250	4350	4125	3900	3700	3475
2.275	4425	4200	4000	3800	3575
2.300	4500	4300	4075	3900	3675
2.325	4575	4375	4175	4000	3800
2.350	4650	4475	4275	4100	3900
2.375	4750	4550	4350	4200	4000
2.400	4825	4650	4450	4300	4125
2.425	4900	4750	4550	4400	4225
2.450	5000	4825	4675	4500	4350
2.475		4900	4750	4600	4450
2.500		5000	4825	4700	4575

For more information please refer to ASTM C 803.

STANDARD TEST METHOD FOR DETERMINING FLOOR FLATNESS AND LEVELNESS USING THE F-NUMBER SYSTEM

This is an Introduction to the new F-Number testing system and is designed for Inspectors who have previous experience performing this type of test as well as Inspectors without the necessary experience. However, it is strongly recommended that the less experienced Inspectors perform this type of testing at least once or twice with an experienced Inspector so they can have quality hands-on experience as well as the opportunity to get their specific technical questions correctly answered. This way most of the questions and/or concerns can be properly addressed.

This material will cover the basic procedures for performing the F-Number Tests.

In order to get familiar with this Code (Standard), you need to refer to the *ASTM E1155-87* and *ACI #117*. This material is based on above mentioned Standards as well as the manufacturing hand book for this Test Method *(FACE Hand book for Model 1600)*.

1. Referenced Documents:

a) ASTM Standard: E 1155-87 & E 1155 M
b) ACI Publications #117
c) FACE Dipstick Hardware Manual for Model 1600

2. What are F-Numbers:

a) The F-Number System is the new **American Concrete Institute** (ACI) standard for specification and measurement of concrete floor flatness and levelness. F-Numbers replace the familiar "1/8th inch in ten feet" type specs that had proven unreliable, unmeasureable and unrealistic.

b) The new standards include two F-Numbers:

Ff for **flatness**   and   Fl for **levelness**

c) **Flatness** relates to the waviness (bumpiness) of the floor, while **Levelness** describes the tilt or pitch of the slab; in other words, it describes the difference in elevation of two points ten feet apart. The higher the F-Number, the more superior the surface of the floor.

d) F-Numbers are linear, thus an Ff 20 is twice as flat as a Ff 10, but only half as flat as a Ff 40.

e) Slabs-on-Grade (SOG) are usually specified with an Ff number and an Fl number (the Ff is always listed first), such as: Ff 25 / Fl 20

f) Because of the deflection, **elevated slabs are usually specified using only Ff.**

g) When a floor is described as an "F 25", it usually means "Ff 25".

h) The F-Number System applies to 99% of all floor slabs-all floors that support *random traffic*, be it vehicular or pedestrian traffic.

i) In the tiny percentage of floors that have **defined traffic**, where vehicles are restricted in their movement by wire or rail guidance, **a different F-Number — F [min]** — is used. This System is used in conjunction with consultation services provided by the FACE Companies.

j) Most **Superflat** floors should use the **F [min]** System, since most of these slabs support **defined traffic.**

3. How are F-Numbers better:

a) F-Numbers control both the floor's "envelope" *and* its bumpiness.

b) Or, if you think of the floor profile as a wave, F-Numbers control both the wave's amplitude and its frequency.

c) F-Numbers have shown the ability to identify and to control floor characteristics which are critical to the floor's usefulness.

4. When should floors be measured:

a) It is recommended that all slabs should be measured within 48 hours after placement of concrete. However, per ACI 117, Fl Numbers shall be measured within 72 hours; there is no time limit for measuring Ff Numbers.

5. How long does it take to measure a floor:

a) A single operator with a Dipstick device can collect enough readings in 60 to 90 minutes to measure the typical day's slab-on-Grade placement. After data collection, an analysis of the readings and a report can be generated on an office PC in about an hour.

6. How do F-Numbers work on elevated slabs:

a) In most cases, only the Ff number is specified on elevated slabs. That is because elevated slabs deflect-and the contractor can't totally control how much deflection occurs.

b) The use of Fl number on elevated slabs is limited to specific situations where the floor profile is analyzed when:

the slab is still supported in its original as-cast position; and

the slab has no camber.

c) It is recommended that no elevated slab should be specified lower than Overall Ff 20, however, refer to the approved plans & specs and if necessary, contact the Architect or Engineer of record prior to the test performance to get clarification in that specific matter.

7. What is a Superflat Floor:

a) By definition, Superflat = F [min] 100

b) The word *Superflat* was created to describe the flatness/levelness required to support full-speed, trouble-free operation of Very Narrow Aisle (VNA) lift trucks.

c) Since F [min] is the Defined Traffic F-Number, *it cannot be measured using the FACE Company's DIPSTICK. This will require a special equipment called Profilograpf.*

d) In order to achieve this high degree of flatness/levelness, Superflat floors must be placed in long, narrow strips. The forms for Superflat placements must be set with great precision. Specialized finishing techniques and continuous quality control measurement are also required. The service of a floor design/construction Consultant are usually advisable.

e) For **Random Traffic Floors**, Superflat is often defined as *Ff 100/Fl 50*; but Superflat tolerances are rarely required on Random Traffic Floors and should be specified only in extraordinary circumstances.

8. Preparation and Procedures for the Test:

Layout:

a) Get the dimensions of the test area. It is recommended that the test area has a rectangular shape, since this way will be easier for performing the test.

b) Bring in the collected dimensions for the test area so that a Layout can be prepared with the help of the "1155 Floor Layout Helper" software.

c) The Layout can be done in couple of ways:

1) Lines add to both lengths and to longest boundary

2) All lines at 45 degree to longest boundary

d) You can choose between these two, however, if there are any permanent items sticking out of the slab, it is strongly recommended that you go with "1" to save yourselves and everyone involved in this matter a headache.

e) Once the layout ha been printed out, you're ready to rock & roll. Take the box with the device with you to the job site along with the layout sheet, tape measure, Marking Crayon, chalk and chalk box. NOTE: Ask the concrete cutters, if applicable, as to what color of chalk will they be using, because you DO NOT want to use the same color chalk as they use.

f) The other thing that you should know about the Layout is, according to ASTM E 1155, "Sample measurement line shall consist of any straight line on the test surface satisfying the following criteria:

1) No sample measurement line shall measure less than eleven (11) feet in length

2) No portion of any sample measurement line shall fall within two (2) feet of any boundary, wall, penetration, or similar disconti-

nuity. [Sample measurement lines on concrete floors may cross shrinkage crack control joints.]"

g) With regards to "2", if you are not using the print out of the above mentioned software, which deducts automatically 2 feet from either side of the given dimensions, please note, in order to perform a valid test in accordance with the standards, you MUST take deduction of 2 feet from either direction of your rectangular test area.

h) Lay out the test area according to the "1155 Floor Layout Helper" print out and/or the specifications that are mentioned above.

i) Divide the entire test surface to test sections. Assign a different ID number to each test section and record the location number to each test section boundaries.

j) Mark each sample measurement line on the test surface. Assign and Record a different ID number to each measurement line.

k) Sample measurement lines within each test section should be arranged either:

Orienting all lines at 45 degree to the longest construction joint abutting the test section, or Placing lines of equal aggregate length both parallel to and perpendicular to the longest test section boundary.

l) **Once you are finished with laying out the test surface, proceed as follow:**

Setup:

a) Open the Dipstick device box,

b) Take out the Dipstick unit; this unit MUST always be calibrated always before performance of the test. Calibration: Install one of the handles in the appropriate place on the Dipstick unit. Stand up the unit on its swivel feet. Turn on the unit by pulling out the switch and pushing it up. Take your marker and mark one of the swivel feet on the surface. Use this mark as your reference point. Look through the both LCD openings. The numbers on both the LCDs should be the same. Remember that number. Now, take the unit off the floor and stand the feet in opposite direction from where you had it before and read the number on both the LCDs. This needs to be done couple of times or until you read the same number by rotating it, whereby you can use the adjust screws on the swivel feet.

c) Once calibrated you are ready to perform the test as required. At this time please turn off the Dipstick unit by pulling out and pushing it down.

d) Take out and install the Pocket Computer in the appropriate place. DO NOT turn the computer on yet. Also, please note: *The pocket computer is the last unit to be turned on and the first unit to be turned off.*

e) Take out and install the computer clamp through the handle on top of the pocket computer so that the cord hanging from the clamp is on the same side as the plug on the Dipstick unit.

f) To connect the beeper, plug the cord from the clamp into the outlet.

g) Now you can install the other two handles on top of the first one.

h) At this time you're finished with setting up the Dipstick device.

Test:

a) Turn on the Dipstick unit as described before. Now push the "ON" button on the pocket computer. Wait for a few seconds until you see the screen giving you 2 options:

  1. Auto Read

  2. Trigger

b) Push "1" for Auto read on the computer.

c) Now the computer shows "Data Menu"

d) Push "F" for files. Now you see the "File Menu"

e) Push "T" to trash the old file. It will ask you if you want to trash all files. Push "Y" to start the process to delete everything.

f) Now the computer will ask you for a new file name. Your file name can have up to eight (8) characters (numbers and/or letters). I recommend a maximum of seven (7) characters; in case of an error you can add another character to your file name; that way you won't lose your data that you've already saved. [example: WSCTC]

g) Then, it will ask you for a date. Dial in only numbers without any spaces, slashes etc. [example: February 5, 2000 = 020500]

h) Now the computer will ask you for the Ff & Fl numbers. For Ff=25 and for Fl=30. These are the overall standards that are recommended by ACI #117.

i) Now you are back in the Data Menu again.

j) Push "F" for files again.

k) Push "N" for new file; these file are sub titled to the File that is already saved.

l) Now it will ask you for the File name. Type the file name [example: WSCTC01]

m) It will ask you whether you want to change the date. Now just push the "Enter" button.

n) Do the same thing with the comments; just push the "Enter" button.

o) Now the computer screen will show:

> *Start data collection*
>
> X = Exit; Enter to begin

p) Stand up the unit on its swivel feet, the word "Start" on the device is the front side of the device and has to be always toward the direction of your measurement line.

q) Once you're ready, push the "Enter" and wait to hear a beep from the device. Once you've heard the beep signal, then you rotate the device in the travel direction [forwards].

r) You do this until the end of each measurement line. Once you are done with collecting data on one line, Push the "!" key.

s) The screen will show you:

FILES HAVE BEEN SAVED!

Turn off computer now, or type RUN to restart.

t) At this time, if you have still other lines for collecting data, type "RUN" and push "Enter" to restart. Follow the same procedure as mentioned above. If you re done with collecting data just turn off the computer and then the Dipstick unit.

u) Disassemble the device and take the pocket computer to your laboratory manager as soon as possible. You have only 24 hours to get the results and prepare the report that needs to be distributed to the parties involved in that particular project.

Report:

a) Write your report as usual; Indicate in your report the area of test [e.g. Grid lines etc.].

EXTERIOR INSULATION AND FINISH SYSTEM (E.I.F.S.) INSPECTION

General:

EIFS is a multi-layer, composite wall cladding system. EIFS is applied by hand in a series of layers. It is applied mostly by plasterers; however, it may be installed by experienced masons, drywallers, and people from other trades, who are experienced.

EIFS is applied in the following three orders:

• Attaching the insulation layer to the wall

• Applying the base coat to the insulation

• Applying the finish to the base coat.

EIFS is a multifunction product that provides three basic functions to the walls:

1) Thermal Insulation

2) Weatherproofing

3) Aesthetics

All three of the above referenced functions are achieved simultaneously by EIFS, which means, if any one of the three EIFS layers is not applied properly, all three functions may be affected. Thus, it is very important and critical that the EIFS installation to be performed carefully.

Tools needed for EIFS Inspection:

- Tape Measure: for general measuring
- Vernier Calipers: for precise measuring of coat thickness, mesh diameter, etc.
- Gallon-size Freezer Bags: for keeping samples of insulation and reinforcing mesh
- 3X5"-Note Cards: for putting notes in freezer bags and for placing next to areas being photographed
- Camera: to document raw materials, problems, etc.
- Permanent Ink Markers: for marking samples
- Metal Corner Square: for checking squareness of insulation edges
- Mason Jars with Caps and Gaskets: for keeping samples of EIFS finish and adhesives, as well as Portland cement
- Scale (Range 0-60 oz): for checking density of insulation boards
- Thermometer: for checking air and substrate temperatures
- Small Mirror on the end of a Small Telescoping Rod: for holding under the bottom edge of EIFS at grade, decks, etc.
- Report forms, Clipboards, and Note Pads
- Hard Hat, Body Harness, Ropes, and Safety Glasses

Below are the basics steps for EIFS inspection. For more detailed description of each of these steps, you may refer to the E.F.I.S. publication for new construction.

- ❖ Review Project Plans and Specifications
- ❖ Meet with owner's representative to determine the level and frequency of inspection to be performed, as well as the type of inspection
- ❖ Identify and follow applicable inspection steps

❖   Write a report to include the following for each site visit:
   a) Project Name
   b) Project Address
   c) Permit Number
   d) Inspector's Name
   e) Client's Name
   f) Site Condition
   g) Substrates
   h) Insulation
   i) Base Coat
   j) Finish
   k) Non-EIFS Wall Components

Levels of Inspection:

A level of inspection is the degree of detail to which the inspection is per-formed. Each EIFS project requires different degrees (or levels) of inspection. However, it is essential for the inspector to understand and ascertain the level of quality, which is appropriate for a given project.

Below are the three levels of inspection:

A)   Level 1: Minimum

This level is the least amount of inspection. This level does not assume that there are any detailed construction documents available. This level also does not require any detailed construction documents be present, it assumes that the inspector is knowledgeable in respect to how EIFS is normally being installed. This level assumes that the manufacturer of EIFS materials will provide basic documentation that may be used by an inspector as the means for determining that the work is being performed correctly. This includes standard construction details and application instructions.

B)   Level 2: Recommended

This level is the most appropriate level for commercial EIFS applications, as well as for custom homes. This level will also cover the substrate and non-EIFS wall components, such as sealants, etc. This level assumes that in addition to the usual documents, such as basic drawings, and manufacturer's provided basic documentation, specific drawings and specifications are created by a design professional.

C)   Level 3: Optional

This level is most commonly used, when the owner is concerned that certain aspects of EIFS application are being performed correctly. For example, the owner may be extremely image conscious and may want to make sure that the color and texture exactly matches a sample that has been provided to her/him. Thus the owner may want the aesthetics of the EIFS finish to be inspected.

Aesthetic vs. Functional Inspections:

Aesthetic inspections are those that involve primarily visual items, which do not affect the durability of the wall. Functional inspections are those that deal with the durability of the wall system. This includes issues such as longevity and water penetration resistance.

Examples of functional inspections include Base Coat thickness, attachment strength and sealant joints. Examples of aesthetic inspection include color/texture assessments, and the straightness of edges.

GUIDELINES FOR HANDLING TEST SAMPLES

A) Sample Receiving:

When samples are received in the laboratory, the samples are placed in the sample receiving area. The sample containers are then labeled with the following information:

1. Project Name

2. Project Number

3. Date Sampled

4. Required Tests

5. Container Number out of total containers per sample

A copy of the field report is placed in the paperwork bin labeled "Sample Paperwork for Laboratory Testing". In addition to the information listed above, the field report will include the sample location and a note identifying the sample as native or imported (for soil samples only).

B) Sample Log-in:

Samples will be logged into the Sample ID Log by a laboratory technician prior to testing. The Sample ID Number, which is a unique sequential number assigned to the samples, will identify the samples.

C) Storage and Disposal:

Samples will remain in the sample receiving area until they are tested with the exception of concrete and masonry samples. Additional procedures for handling concrete and masonry samples are contained in separate procedure. Once a sample is tested and the report is issued, the sample will be discarded within 3-4 days.

CLASSIFICATION OF SOILS FOR ENGINEERING PURPOSES

A) All soils received into the laboratory will be described and classified by the Unified Soil Classification System (ASTM D2487-93) as modified by the following guidelines.

B) Terminology:

Gravel: Particles passing a 3" sieve and retained on a Number 4 sieve

Coarse Gravel: Majority of particles retained on the 3/4" sieve

Fine Gravel: Majority of particles passing the 3/4" sieve

Sand: Particles passing a No. 4 sieve and retained on a No. 200 sieve

Coarse Sand: Majority of particles retained on the No. 10 sieve

Medium Sand: Majority of particles passing No. 10 and retained on No. 40 sieve

Fine Sand: Majority of particles passing No. 40 sieve

Fines: Particles passing No. 200 sieve

Silt Fines: Particles less than .005mm and greater than .002mm (typically non-plastic)

Clay Fines: Particles less than .002mm, typically plastic

109

C) Soil Descriptions:

Soil is described as outlined in ASTM D2487. The primary constituent is named first. Secondary constituents are further enumerated with the following terminology:

Trace (trc): This particle size makes up only 5-10% of overall material gradation

Some (sm): This particle size makes up 10-20% of overall material gradation

Suffix y (e. g. sandy): This particle size makes up 30-40% of overall material gradation

And (&): This particle makes up 40-50% of overall material gradation.

GUIDELINES FOR SAMPLING AGGREGATES

A) General:

For sampling aggregates follow the following guideline and for references follow the ASTM D75.

B) Procedure:

When sampling from bins or moving conveyor belts, the aggregate shall be sampled in a way that 3 samples are taken from the entire cross section of the discharge location. All 3 samples are combined to form one field sample.

When sampling from stopped conveyor belts, obtain at least three samples from the belt. Once the belt is stopped, two templates are placed on the belt and all material in between the two templates will be collected. This procedure shall be done in three locations. All three samples are combined to build a uniform field sample.

Sampling from stock piles shall be avoided wherever possible. If a stock pile must be sampled, then three samples must be obtained: one from the top third, one from the middle third, and one from the bottom third. These samples are combined to form a uniform field sample.

Samples from roadways (bases and sub-bases) shall be obtained from at least three different locations. Precautions should be taken not to obtain any of the underlying material. The three samples are then combined to form an uniform field sample.

The size of aggregate samples shall be a minimum of the following:

Maximum Nominal Size	Min. Mass of Sample per lbs (kg)	
No. 8-3/8"	25	(10)
1/2 inch	35	(15)
3/4 inch	55	(25)
1 inch	110	(50)
1-1/2 inch	165	(75)

C) Tractability and Handling of Samples:

The sample is placed into a cloth or plastic bag using the appropriate sampling methods as described above. The bag is then sealed and tagged for identification, tests to perform, and transported to the laboratory. When a sample is brought into the laboratory, the aggregate is placed in the designated "Sample Receiving Area". All special instructions shall be written on the field report that identifies the sample.

A copy of the field report must be left on the sample, and the original report is given to the laboratory manager who will log the sample in the Sample ID book and assign it an ID number. A copy of report with assigned ID number must be returned to laboratory with sample. Original report is kept in the lab until laboratory work is completed. The date that the sample was logged in, is recorded as well as the project number, project name, tests to perform, any special notes, date completed and the name of the technician, who performed the tests.

D) Reducing Field Samples to Appropriate Test Sizes:

Samples will be reduced to testing size by quartering method, or use of quartering cloth.

Organic Impurities in Fine Aggregates For Concrete

A) This method is performed based upon the following procedure, which reflects the requirements prescribed by ASTM C40.

B) Procedure:

 After the sample is obtained, it shall be split down to a one lb. size. Next, the reagent is made by mixing 3 parts reagent grade sodium hydroxide (NaOH) to 93 parts clean water. Next, a glass bottle is then filled to the 4.5 fluid oz level with the fine aggregate. The bottle is then filled to the 7 fluid oz line with the 3% NaOH solution and the bottle is shaken to release air voids. When this is filled to the 7 fluid oz mark, the bottle is sealed and shaken vigorously until thoroughly mixed. The bottle is then allowed to sit for 24 hours. Then, the test solution is compared to the standard color chart and the color plate number is noted. If the test sample is darker than the standard, the aggregate under test shall be considered to possibly contain injurious organic impurities. The test results are then recorded and distributed.

MATERIALS FINER THAN NO. 200 SIEVE IN MINERAL AGGREGATE BY WASHING

A) This method is performed based upon the following procedure, which reflects the requirements prescribed by ASTM C117.

C) Procedure

After a field sample has been taken the sample is then split down to the appropriate size.

Nominal Maximum Size		Minimum Mass g
2.36mm	(No. 8)	100
4.75mm	(No. 4)	500
9.5mm	(3/8 in)	1000
19mm	(3/4 in)	2500
37.5mm	(1.5 in +)	5000

The appropriate sized sample is dried to a constant mass at 230 + 9oF. The mass is then determined to the nearest 0.1 gram. The sample is then covered with water and then agitated to completely separate finer particles. Then the wash is poured over the No. 200 sieve being sure not to spill any material. This is continued until the wash water appears clear. Wash material retained on the No. 200 sieve back into the container. Then the material is dried at 230 + 9oF until at a constant mass.

The following calculation determines the percent passing the No. 200 sieve.

a) mass of original dry sample
b) mass of dry sample after wet wash

finer than No. 200.%=[(a-b)/a]x100

The results are then recorded and distributed.

METHOD OF TEST FOR DETERMINATION OF DEGRADATION VALUE

A) **Scope:** This method covers procedures for determining the quality of fines produced by self-abrasion of aggregate in the presence of water.

B) **Apparatus:**

1. Balance – 2000 g capacity, sensitive to 1 g
2. Sieve shaker – with 44.45 mm (1 ¾ in.) throw on cam at 300 +/- 5 oscillations per minute.
3. Plastic canister – 190.5 mm (7 ½ in.) diameter x 152.4 mm (6 in.) high
4. Sand equivalent graduated cylinders
5. Sand equivalent stock solutions
6. Sieves – 2.00 mm (U.S. No. 10) and 0.075 mm (U.S. No. 200) sieves
7. Graduates – 500 ml tall form, 10 ml
8. Interval timer

C) **Procedure:**

1. Crush the material to be tested to pass the 12.5 mm sieve (1/2 in.), wash over a 2.00 mm sieve (U.S. No.10) and dry to constant weight

2. 1000 g sample of the aggregate graded as follows:

12.5 mm (1/2 in.) – 6.3 mm (1/4 in.)	500 g
6.3 mm – 2.00 mm (1/4 in.) sieve (U.S. No. 10)	500 g

3. Place sample in the plastic canister, add 200 cc of water, cover tightly and place in sieve shaker

4. Agitate the material for 20 minutes

5. Empty the canister into nested 2.00 mm (U.S. No. 10) and 0.075 mm (U.S. No. 200) sieves placed in a funnel over a 500 ml graduate to catch all the water

6. Wash out the canister and continue to wash the aggregate with fresh water until wash water in the graduate is filled to the 500 ml mark. (the aggregate may drain 50 to 100 ml of water after washing has been stopped.)

7. Pour 7 ml of sand equivalent stock solution into a sand equivalent cylinder

8. Bring all solids in the graduate into suspension by capping the graduate with the palm of the hand and turning it upside down and back as rapidly as possible for about 10 minutes

9. Immediately decant into the sand equivalent cylinder to the 381 mm (15 in.) mark and insert stopper in the cylinder

10. Mix the contents of the cylinder by alternately turning the cylinder upside down and right side up, allowing the bubble to traverse from end to end. Repeat this cycle 20 times in approximately 35 seconds

11. Place the cylinder on the table, remove stopper, and start timer. After 20 minutes read and record the height of the sediment column to the nearest 2.0 mm (01. In.).

D) Calculation:

1. Calculate the degradation factor by the following formula:

$$D = \frac{(15 - H)}{(15 + 1.75\,H)} \times 100$$

Where:

D = Degradation Factor

H = Height of Sediment in Tube

3. Values may range from 0 to 100, with high values being best materials. The formula places doubtful materials at about the midpoint of the scale, with poor ones below, and good ones above the point.

Table 1: Degradation Value "D"

H	D	H	D	H	D	H	D	H	D
0.0	100	3.1	58	6.1	35	9.1	19	12.1	8
0.1	98	3.2	57	6.2	34	9.2	19	12.2	8
0.2	96	3.3	56	6.3	33	9.3	18	12.3	7
0.3	95	3.4	55	6.4	33	9.4	18	12.4	7
0.4	93	3.5	54	6.5	32	9.5	17	12.5	7
0.5	91	3.6	54	6.6	32	9.6	17	12.6	6
0.6	90	3.7	53	6.7	31	9.7	17	12.7	6
0.7	88	3.8	52	6.8	30	9.8	16	12.8	6
0.8	87	3.9	51	6.9	30	9.9	16	12.9	6
0.9	85	4.0	50	7.0	29	10.0	15	13.0	5
1.0	84								
1.1	82	4.1	49	7.1	29	10.1	15	13.1	5
1.2	81	4.2	48	7.2	28	10.2	15	13.2	5
1.3	79	4.3	48	7.3	28	10.3	14	13.3	4
1.4	78	4.4	47	7.4	27	10.4	14	13.4	4
1.5	77	4.5	46	7.5	27	10.5	13	13.5	4
1.6	75	4.6	45	7.6	26	10.6	13	13.6	4
1.7	74	4.7	44	7.7	26	10.7	13	13.7	3
1.8	73	4.8	44	7.8	25	10.8	12	13.8	3
1.9	71	4.9	43	7.9	25	10.9	12	13.9	3
2.0	70	5.0	42	8.0	24	11.0	12	14.0	3
2.1	69	5.1	41	8.1	24	11.1	11	14.1	2
2.2	68	5.2	41	8.2	23	11.2	11	14.2	2
2.3	67	5.3	40	8.3	23	11.3	11	14.3	2
2.4	66	5.4	39	8.4	22	11.4	10	14.4	1
2.5	65	5.5	39	8.5	22	11.5	10	14.5	1
2.6	63	5.6	38	8.6	21	11.6	10	14.6	1
2.7	62	5.7	37	8.7	21	11.7	9	14.7	1
2.8	61	5.8	37	8.8	20	11.8	9	14.8	0
2.9	60	5.9	36	8.9	20	11.9	9	14.9	0
3.0	59	6.0	35	9.0	20	12.0	8	15.0	0

STANDARD TEST METHOD FOR REBOUND NUMBER OF HARDENED CONCRETE

1. General:

This test method will cover the determination of a rebound number of hardened concrete using a spring-driven steel hammer, such as SCHMIDT HAMMER.

Schmidt Hammer will impact with a predetermined amount of energy, a steel plunger in contact with a surface of concrete, and the distance that the hammer rebounds is measured. This test method can be used to assess the in-place uniformity of concrete, to delineate regions in a structure of poor quality or deteriorated concrete, and to estimate In-place strength of concrete members.

If you are using this test method to estimate strength, it requires to establish a relationship between strength and rebound number. The relationship should be established for a given concrete mixture and given apparatus. This relationship should also be established over the range of concrete strength that is in question. For estimating the strength during construction, establish the relationship by performing rebound number tests on molded specimens and measuring the strength of the same or companion specimens. To estimate the strength in an existing structure, establish the relationship by correlating rebound numbers measured on the structure with the strengths of cores taken and tested from corresponding locations. Also, refer to ACI 228.1R for additional information regarding development of relationship and using the relationship to estimate in-place strength.

For any concrete mixture, the rebound number is affected by many factors, such as moisture content of the test surface, the method used to obtain the test surface (such as types of finishes, or form materials), and the depth of carbonation. These factors must be considered in preparing the strength relationship and interpreting test results.

The Schmidt hammer consists of a spring-loaded steel hammer which, when it is released, strikes a steel plunger in contact with the concrete surface. The spring-loaded hammer must travel with a consistent and reproducible velocity (force). The rebound distance of the steel hammer from the steel plunger is measured on a linear scale attached to the frame of the Schmidt Hammer.

2. Testing Area:

Concrete members to be tested should be at least 4 inches thick and fixed within the structure. Avoid areas that exhibit honeycombing, scaling, or high porosity. As a ground rule, troweled surfaces generally exhibit higher rebound number than screed or formed finishes. Where possible, structural slabs should be tested from the underside to avoid finished surfaces.

The test area should be at least 6 inches in diameter. Surfaces with loose mortar, soft, or heavily textured surfaces must be ground smooth with the abrasive stone, which consist of medium-grain texture silicon carbide or equivalent materials. Troweled or smooth-formed surfaces need not be ground prior to testing. However, please note that, where formed surfaces are ground, it increases in rebound number of 0.4 for high-density plywood formed surfaces and 2.1 for plywood formed surfaces. Also, wet concrete surfaces give lower rebound numbers than dry surfaces. Presence of concrete surface carbonation may result in higher rebound numbers. Dry surfaces and surface carbonation will result in higher rebound numbers. However, this can be reduced by thoroughly wetting the surface for

24 h prior to testing. If the carbonated concrete surface is very thick, it may be necessary to remove the carbonated layer in the test area, utilizing a power grinder. This way we can obtain rebound numbers that are representative of the interior concrete.

Never compare the ground and ungrounded surfaces.

Other factors that may affect the results of the test are:

- Temperature of concrete: concrete at 32 degree F or less, may exhibit a higher rebound numbers. Concrete surface shall be tested only after it has thawed.
- Temperature of the Schmidt Hammer itself may affect the rebound values.
- Different hammers of the same nominal design may give rebound numbers differing from 1 to 3 units and therefore, test should be made with the same hammer in order to compare results.
- Rebound hammers shall be serviced and verified semiannually and whenever there is reason to question their proper operation.

3. Test Procedure or how to use the Schmidt Hammer:

Hold the Schmidt Hammer firmly so that the plunger is perpendicular to the test surface. Gradually push the instrument toward the test surface until the hammer impacts. After impact, maintain pressure on the instrument and, if necessary, depress the button on the side of the instrument to lock the plunger in its retracted position. Estimate the rebound number on the scale to the nearest whole number and record the rebound number. Take at least ten readings from each test area. No two impact-tests should be closer than 1 inch. Examine the impression made on the surface after impact, and if the impact crushes or breaks through a near-surface air void disregard the reading and take another reading.

4. Calculation

Discard readings differing from the average of 10 readings by more than 6 units and determine the average of the remaining readings. If more than two readings differ from the average by 6 units, discard the entire set of readings and determine rebound numbers at 10 new locations within the test area.

5. Report:

Report the following information for each test area:

    a) Date and time of testing

    b) Identification of location tested in the concrete construction and the type and size of member tested

    c) Description of the concrete mixture proportions including type of coarse aggregate, if known

    d) Design strength of concrete tested

    e) Description of the test area including surface characteristics, etc.

    f) If surface was ground and depth of grinding

    g) Type of form material used for test area

    h) Curing conditions of test area

    i) Type of exposure to the environment

    j) Hammer identification and serial number

    k) Air temperature at the time of testing

    l) Orientation of hammer during test

    m) Average rebound number for each test area

    n) Remarks regarding discarded readings of test data or any unusual conditions or findings.

TURBIDITY MONITORING REQUIREMENTS

1. Definition:

Turbidity is a measure of the extend to which light is either absorbed or scattered by suspended material in water. Because both size and surface characteristics of the suspended material influence absorption and scattering, turbidity is not a direct quantitative measurement of suspended solids. The level of turbidity is determined by measuring the amount of light that is scattered as it passes through a standard sample of the water. This measurement is expressed in Nephelometric Turbidity Units (NTU).

2. Applicable Codes

Each Jurisdiction has its own codes requirements, which can be found under the clearing and grading code. In State of Washington, all the local codes are written in compliance with the State Surface Water Quality Standards (WAC 173.201A-030).

For example, the above referenced Washington State Code requires the following standards for turbidity:

- It is not to exceed 5 NTU over upstream turbidity when upstream turbidity is 50 NTU or less;
- It is not to exceed 10% above upstream turbidity when upstream turbidity is greater than 50 NTU.

3. Turbidity Monitoring Plan Requirements:

a) *Project Description:* This section of the plan should identify the purpose of the site clearing and grading, including a discussion of the extent of site disturbance required for the proposal, any proposed phasing of the project, and a brief description of the Temporary Erosion Control Plan (TESC Plan).

b) *Drainage Analysis:* This section, at a minimum, should include a discussion of the general topography; existing drainage patterns on-site including existing drainage features (i.e. wetlands, streams, ditches, catch basins, pipes, ponds, etc.); and location of protected areas (i.e. steep slopes, wetlands, riparian corridors and shorelines).

c) *Monitoring Locations:* All upstream and downstream monitoring points must be indicated in the turbidity monitoring plan and on the TESC Plan. If more than one drainage basin (sub-basin) is present, multiple upstream and downstream monitoring points will be required to accurately monitor the site. Monitoring locations should be immediately downstream from discharge point.

Please note that when measuring turbidity from point source discharges (e.g. releasing water from a sediment pond into the local drainage system); avoid sampling discharge prior to its mixing with local stormwater. For example, when discharging to a catch basin, take the sample from the inlet of the next catch basin downstream, thus allowing mixing of discharge and off-site stormwater.

d) *Gathering Baseline Data:* Baseline data is used in lieu of upstream data, when upstream data is not attainable during construction. Baseline data must be gathered prior to initiating clearing and grading or other site disturbance (i.e. demolition, etc.). To gather data, the inspector shall sample turbidity at each of the downstream monitoring points at least two times per week for a period not less than two weeks. The baseline turbidity for each monitoring point is defined as the average turbidity of the samples taken at that location.

e) *Turbidity Monitoring Data Sheet:* The data sheet should have at least the following information:

- Project Name
- Project Address
- Permit Number
- Inspector's Name
- Date & Time of Sample
- Baseline Turbidity in NTU
- Weather Conditions
- Downstream & Upstream location and their Reading NTU's
- Whether turbidity increases upstream or downstream (NTU)
- Allowable Turbidity Increase (NTU)
- Contractor notification and result
- Corrective Measures Taken by Contractor
- Other Comments

f) *Field Testing Method:* Specify turbidity monitoring equipment used, which shall comply with the requirements of the EPA.

g) *Frequency of Monitoring:* During the dry season (typically May – October) sampling shall be performed no less than one sample weekly. Additional samples shall be taken during each rainfall event. No more than one sample will be required, if in any given test-day, test indicates that turbidity complies with allowable levels. If the test indicates that turbidity is in excess of the standard or turbid water is observed coming from the site after the initial sample is taken, additional samples may be required. Sampling during the rainy season (typically November – April) shall be done daily, preferably during rainfall events.

h) *Determination of Compliance:* To determine turbidity level compliance, upstream turbidity data is subtracted from the downstream data. The increase in turbidity is then compared to the applicable codes and requirements (in this case to the State Surface Water Quality Standards – WAC 173.201A-030). A negative increase indicates that water from the site is cleaner than the water upstream, and no correction in necessary.

i) *Reporting Requirements:* Sampling data sheets shall be delivered to each jurisdiction the same day they are taken.

FUNDAMENTALS OF BRIDGE INSPECTION

1. *Introduction:* Bridge inspection is playing an increasingly important role in providing a safe infrastructure of our nation. As our nation's bridges continue to age and deteriorate, an accurate and thorough assessment of each bridge's condition is critical in maintaining a trustworthy highway system.

This chapter will present the responsibilities and duties of the bridge inspector.

2. *Responsibilities and Duties of the Bridge Inspector:* There are five basic responsibilities of the bridge inspector:
   - Maintain public safety and confidence
   - Protect public investment
   - Provide bridge inspection program support
   - Provide accurate and reliable bridge records
   - Fulfill legal responsibilities

The primary responsibility of the bridge inspector is to maintain public safety and confidence. The general public travels highways and bridges without hesitation. However, when a bridge fails, the public's confidence in the bridge system is violated. The design engineer's role in assuring bridge safety is:
   - To incorporate safety factors
   - To design conservatively when appropriate

The inspector's role is:

- To provide thorough inspections identifying bridge conditions and defects
- To prepare condition reports documenting these deficiencies and presenting recommendations

Another responsibility is to protect public investment in bridges. The inspector must be on guard for minor problems, which can be corrected before they lead to costly major repairs. The inspector must also be able to recognize bridge elements, which need repair in order to maintain bridge safety and avoid replacement costs.

The National Bridge Inspection Standards (NBIS), part of the Code of Federal Regulations, mandates:

- Inspection procedures
- Frequency of inspections
- Qualifications of personnel
- Reporting
- Inventory

There are three major reasons why accurate and reliable bridge records are required:

- To establish and maintain a structure history file
- To identify and assess bridge repair requirements
- To identify and assess bridge maintenance needs

A bridge inspector report is a legal document. Therefore, descriptions must be concise, specific, detailed, quantitative (where possible), and complete.

Also, original inspection notes should not be altered without approval of the inspector, who wrote the notes. A bridge inspection should be performed in accordance with NBIS, as well as local Department of Transportation requirements.

There are five basic duties of the bridge inspector:
- Planning the inspection
- Preparing for the inspection
- Performing the inspection
- Identifying items for repairs and maintenance
- Preparing, distributing, and filing the report

In order to make the inspection as systematic as possible, the inspector should make plans in advance. These plans should include determining the inspection sequences, establishing a time schedule, preparing for special inspection requirements (such as nondestructive examinations, etc.), organizing the field notes, anticipating the effects of traffic control procedures, and any other measures to facilitate a thorough and complete inspection.

Preparation measures needed prior to the inspection include organizing proper tools and equipment, reviewing the bridge structure files, and locating plans for structure.

Duties associated with the inspection include maintaining the proper structure orientation and member numbering system, developing an inspection sequence, and following proper inspection procedures. Also, documentation is essential for an in-depth inspection. The inspector must gather enough information to ensure a comprehensive and complete report.

The final basic duty of a bridge inspector is to identify items for repairs and maintenance. The inspector must identify such items to ensure public safety and maximum longevity of the bridge.

3. *Preparation for Inspection:* The success of the on-site field inspection is largely depending on the effort spent in preparing for the inspection. The major preparation activities are:

    - Reviewing the bridge structure file
    - Identifying the components and elements
    - · Developing an inspection sequence, if not in-place or if it needs modifications
    - Preparing notes, forms, sketches, etc.
    - Arranging for traffic control
    - Accounting for special considerations
    - Organizing the tools and equipment
    - Making arrangements for required methods of access
    - Reviewing safety precautions

    a) *Reviewing the bridge structure file:*

The first step in preparing for a bridge inspection is to review the many available sources of information about the bridge, such as:

- **As-built bridge plans:** The bridge plans contain information about the bridge type, the number of spans, the use of simple or continuous spans, and the materials of construction. They also contain information about the presence of composite action between the deck and girders, the use of framing action at the substructure members, and the kind of connection details used. Also, the year of construction and the design loading are usually contained in the bridge plans.

- **Previous inspection reports:** Previous inspection reports provide valuable information about the history of the bridge, documenting its condition in previous years. This information can be used to determine which components and elements of the bridge need special attention. It also allows the inspector to compare the current levels of deterioration with those noted during previous inspections.

- **Maintenance and repair records:** Maintenance and repair records allow the bridge inspector to report all subsequent repairs during the inspection phase, noting the types, extend, and dates of the repairs.

- **Hydrologic and geotechnical data:** Hydrologic data provides information about the shape and location of the channel, the presence of protection devices, flood frequencies, and water elevations for various flood intervals. Using this information, the inspector can note any changes in the channel configuration and in the water elevation. Geotechnical data provides information about the foundation material below the structure. Sand, silt, or clay is more vulnerable to settlement and scour problems than is rock. Therefore, structures founded on these materials should generally be given more attention than those founded on rock.

- **Utility and Right-of-way plans:** Utility plans can be used to determine the types and numbers of utility attachments, and right-of-way plans can be used to determine the limits of the right-of-way.

b) *Identifying Components and Elements:*

Another important activity in preparing for the inspection is to establish the structure orientation, as well as a system for identifying the various components and elements of the bridge. If drawings or previous inspection reports are available, the identification system used during the inspection should be the same as that used in these sources.

However, if no previous records are available, then the inspector should establish an identification system. The numbering system must be the same as used throughout the state. The route direction can be determined based on mile markers or stationing, and this direction should be used to identify the beginning and the end of the bridge.

The deck element numbering system should include the deck section (between construction joints), expansion joints, railing, parapets, and light standards. These elements should be numbered consecutively, from the beginning to the end of the bridge.

c) *Developing Inspection Sequence:*

An inspection normally begins with the deck and superstructure elements and proceeds to the substructure. However, there are many factors that must be considered when planning a sequence of inspection for a bridge, including:

- Type of bridge
- Condition of the bridge components
- Overall condition
- Inspection agency (e.g. DOT) requirements
- Size and complexity of the bridge
- Traffic conditions
- Special procedures

A sample inspection sequence for a bridge of average length and complexity is given below. While developing an inspection sequence is important, it is of value, if following the sequence ensures a complete inspection of the bridge.

Sample Inspection Sequence:

i) **Roadway and Deck Elements**
- Approach roadways
- Traffic safety features
- Bridge deck
- Expansion joints
- Sidewalks and railings
- Drainage
- Signing
- Electrical/lighting
- Barriers, gates, and other traffic control devices

ii) **Superstructure Elements**
- Bearings
- Main supporting members
- Secondary members and bracings
- Utilities
- Anchorages

iii) **Substructure Elements**
- Abutments
- Skewbacks (arches)
- Slope protection
- Piers
- Footings
- Piles
- Curtain walls

iv) **Channel and Waterway Elements**

- Channel profile and alignment
- Channel streambed
- Channel embankment
- Channel embankment protection
- Fenders
- Dolphins
- Hydraulic opening
- Water depth scales
- Navigational lights and aids

d. *Preparing Notes*

Preparing notes, forms and sketches prior to the on-site inspection elimi-nates unnecessary work in the field. Copies of the DOT standard inspec-tion forms should be obtained for use in recordkeeping.

Photocopy sketches from previous inspection reports so that defects pre-viously documented can simply be updated. Preparing extra copies pro-vides a contingency for sheets that may be lost or damaged in the field. If previous sketches are not available, pre-made, generic sketches may be used for repetitive features or members. Possible applications of this time-saving procedure include deck sections, floor systems, bracing members, abutments, piers, and retaining walls.

e. *Traffic Control, etc.*

Bridge inspection, like construction and maintenance activities on bridges, often presents motorists with unexpected and unusual situations. Most state agencies have adopted the federal *Manual on Uniform Traffic*

Control Devices for Streets and Highways (MUTCD). Some state and local jurisdictions, however, issue their own manuals. When working in an area exposed to traffic, the bridge inspector should check and follow the existing standards. These standards will prescribe the minimum procedures for a number of typical applications and the proper use of standard traffic control devices, such as cones, signs, and flashing arrow boards.

f. *Special Conditions:*

The inspection report or the bridge record file should state the amount of time required for the inspection. The inspection time requirements should be broken down into office preparation, travel time, field time, and report preparation time. In populated areas, an inspection requiring traffic restrictions may be limited to certain hours of the day, such as 11:00 am to 1:30 pm. Some days may be banned for inspection work altogether. Actual inspection time may be less than a 40-hour work week in these situations, and schedules should be adjusted accordingly.

Set-up time must be considered both before and during the inspection. For example, rigging efforts may require several days before the inspectors arrive on the site. Also, other equipment, such as compressors and cleaning equipment, may require daily set-up time. Adequate time should be provided in the schedule for set-up and take-down time requirements.

Access requirements should also be considered when preparing for an inspection. Bridge members may be very similar to each other, but they may require different amounts of time to gain access to them. For example, it may take longer to maneuver a lifting device to gain access to a floor system near utility lines than for one that is free of obstructions.

The overall condition of the bridge will play a major role in determining how long an inspection will take. Previous inspection reports usually provide an indication of the bridge's overall condition. It usually takes more time to inspect and document findings, such as a deteriorated element, than it does to simply observe that an element is in good condition. Adverse weather conditions may not halt an inspection entirely, but may play a significant role in the inspection process. During these conditions, climbing should generally be avoided. There must be an increased awareness of safety hazards, and keeping notes dry can be difficult. During seasons of poor weather, a less aggressive schedule should be in-place than during good weather months.

When inspecting a bridge owned by or crossing a railroad, an access permit must be obtained before proceeding with the field inspection. A permit must also be obtained when inspecting bridges passing over navigable waterways.

g. *Additional Preparation:*

Additional preparation for inspection includes: organizing tools and equipment, as needed; making arrangements for required methods of access; and reviewing safety precautions.

4. *Inspection Procedures:* The procedures used to inspect a bridge depend largely on the bridge type, the materials used, and the general condition of the bridge. Therefore, the inspector must be familiar with the basic inspection procedures for a wide variety of bridges. A first step in the inspection procedure is to establish the orientation of the site and of the bridge. The orientation should include the compass directions, the direction of waterway flow, and the direction of the inventory route. Numbers

or letters should be crayoned or painted on the bridge to identify and code components and elements of the structure. The purpose of these marks is to keep track of the inspector's location and to guard against overlooking any portion of the structure.

After the site orientation has been established, the inspector is ready to begin the on-site inspection. The inspector must be careful to the work at hand, and no portion of bridge should be overlooked. Those portions that are most critical to the structural integrity of the bridge should be given special attention. The prudence used during the inspection must be combined with thorough and complete recordkeeping. A very careful inspection is worth no more than the records kept during that inspection.

Basic Guidelines for Inspection:

Decks:
The inspector should check the approach pavement for unevenness, settlement, or roughness. Also check the condition of the shoulders, slopes, drainage, and approach guardrails. The deck and any sidewalk should be examined for various defects, noting size, type, extent, and location of each defect. The location should be referenced using the centerline or curb line, the span number, and the distance from a specific pier or joint.

Examine the expansion joints for sufficient clearance and for adequate seal. Record the width of the joint opening at both curb lines, noting the air temperature and the general weather conditions at the time of the inspection.

Finally, check that safety features, signs, and lighting are present and identify their condition.

Superstructures:

The superstructure must be inspected thoroughly, since the failure of a main supporting member could result in the collapse of the bridge. The most common forms of main supporting members are:

- Floor beams and stringers
- Trusses
- Beams and girders
- Suspender cables
- Eyebar chains
- Arch ribs
- Frames
- Pins and hanger plates

The bearings must also be inspected thoroughly, since they provide the critical link between the superstructure and the substructure. Record the difference between the rocker tilt and a fixed reference line, noting the direction of tilt, the air temperature, and the general weather conditions at the time of the inspection.

Substructures:

The substructure, which supports the superstructure, is made up of abutments, piers, and bents. If "as-built" plans are available, the dimensions of the substructure units should be compared with those presented on the plans. Since the primary method of bridge inspection is visual, all dirt, leaves, animal waste, and debris should be removed to allow close observation and evaluation. Substructure units should be checked for settlement by sighting along the superstructure and plumbing vertical faces. In conjunction with the Scour Inspection of the waterway, the substructure units should be checked for undermining, noting both its extent and location.

Waterways:

Waterways are dynamic in nature, with their volume of flow and their path continually changing. Therefore, bridges passing over them must be carefully inspected for the effects of these changes. A record should be maintained of the channel profile and alignment, noting any meandering of the channel both upstream and downstream. Report any skew or improper location of the piers or abutments.

Scour is the primary concern when evaluating the effects of waterway on bridges. The existence and extent of scour must be determined using a grid system and noting the depth of the channel bottom at each grid point. Scour Inspection is a chapter by itself. It will be discussed in the next chapter.

Embankment erosion should also be noted both upstream and downstream of the bridge, as should debris and excessive vegetation. Record their type, size, extent, and location. Note also the high water mark, referencing it to a fixed elevation such as the bottom of the superstructure.

Inspection of Bridge Elements:

The inspector must be familiar with several general terms used to describe bridge defects:

- Corrosion – rusting
- Cracking – breaking away without separating into parts
- Splitting – separating into parts
- Connection slippage – connections coming apart
- Overstress – deformation due to overload

- Collision damage – damage caused when a bridge is struck by vehicles or vessels

Concrete Inspection: When inspecting concrete structures, note all visible cracks, recording their type, width, length, and location. Any rust or efflorescence stains should also be recorded. Concrete scaling can occur on any exposed face of the concrete surface, and its area, location, depth, and general characteristics should be recorded. Inspect concrete surfaces for delamination or hollow zones, which are areas of incipient spalling, using a hammer or a chain drag. Delamination should be carefully documented using sketches showing the location and pertinent dimension.

Unlike Delamination, spalling is readily visible. Spalling should also be documented using sketches, noting the depth of the spalling, the presence of exposed reinforcing steel, and any deterioration or section loss that may be present on the exposed bars.

Timber Inspection: When inspecting timber structures, determine the extent and severity of weathering and wear, being specific about dimensions, depths, and locations. Probe the timber to detect any hidden deterioration due to decay, insects, or marine borers.

Note any large cracks, splits, or crushed areas. While these may be caused by collision or overload damage, the inspector should be factual, avoiding speculation as to the causes. Note any fire damage, recording the measurements of remaining sound material. Document any exposed untreated portions of the wood, indicating the type, size, and location.

Steel and Iron Inspection: When inspecting steel or iron structures, determine the extent and severity of corrosion, carefully measuring the amount of cross section loss. All cracks should be noted, recording their length, size, and location. Bent or damaged members should be documented, noting the type of damage and amount of deflection.

Loose rivets or bolts can be detected by striking them with a hammer while holding a thumb on the opposite end of the rivet or bolt. Movement will be felt, if it is loose. In addition, any missing rivets or bolts should also be noted.

Note any frozen pins, hangers, or expansion devices. One indication of this is if the hangers or expansion rockers are inclined or rotated in a direction opposite to that expected for the current temperature. In cold weather, a rocker bearing should lean towards the fixed end of the bridge, while in hot weather, it should lean away from the fixed end. A locked bearing is generally caused by heavy rust on the bearing elements.

INSPECTION OF BRIDGES FOR SCOUR

1. Introduction

There are two main objectives to be accomplished in inspecting bridges for scour:

a) to accurately record the present condition of the bridge and the stream;

b) to identify conditions that are indicative of potential problems with scour and stream stability for further review and evaluation by others.

In order to accomplish these objectives, the inspector needs to recognize and understand the interrelationship between the bridge, the stream, and the floodplain. Typically, a bridge spans the main channel of a stream and perhaps a portion of the floodplain. The road approaches to the bridge are typically on embankments, which obstruct flow on the floodplain. This overbank or floodplain flow must, therefore, return to the stream at the bridge and/or overtop the approach roadways. Where overbank flow is forced to return to the main channel at the bridge, zones of turbulence are established and scour is likely to occur at the bridge abutments. Further, piers and abutments may present obstacles to flood flows in the main channel, creating conditions for local scour because of the turbulence around the foundations. After flowing through the bridge, the floodwater will expand back to the floodplain, creating additional zones of turbulence and scour.

If the bridge is determined to be scour critical, a plan of action should be developed for installing scour countermeasures. In this case, the rating of the bridge substructure should be revised to reflect the effect of the scour on the substructure.

2. Office Review

It is desirable to make an office review of bridge plans and previous inspection reports prior to making the bridge inspection. Information obtained from the office review provides a better basis for inspecting the bridge and the stream. Items for consideration in the office review include:

- Has an engineering scour evaluation study been made? If so, is the bridge scour critical?

- If the bridge is scour critical, has a plan of action been made for monitoring the bridge and/or installing scour countermeasures?

- What do comparisons of streambed cross section taken during successive inspections reveal about the streambed? Is it stable? Degrading? Aggrading? Moving laterally? Are there scour holes around piers and abutments?

- What equipment is needed to obtain streambed cross section?

- Are there sketches and aerial photographs to indicate the planform location of the stream and whether the main channel is changing direction at the bridge?

- What type of bridge foundation was constructed? (spread footings, piles, drilled shafts, etc.) Do the foundations appear to be vulnerable to scour?

- Do special conditions exist requiring particular methods and equipment for underwater inspections?

- Are there special items that should be looked at?

3. Bridge Inspection

As discussed in the last chapter, during the bridge inspection, the condition of the bridge waterway opening, substructure, channel protection, and scour countermeasures should be evaluated, along with the condition of the stream. The 1988 FHWA "Bridge Recording and Coding Guide" contains material for the following three items:

a) Item 60: Substructure

b) Item 61: Channel and Channel Protection

c) Item 71: Waterway Adequacy

The guidance in the "Bridge Recording and Coding Guide" for rating the present condition of items 61 and 71 is set forth in detail. Guidance for rating the present condition of Item 60, Substructure, is general and does not include specific details for scour. The following section presents approaches to evaluating the present condition of the bridge foundation for scour and the overall scour potential at the bridge.

Assessing the Substructure Condition: Item 60, Substructure, is the key item for rating the bridge foundations for vulnerability to scour damage. When a bridge inspector finds that a scour problem has already occurred, it should be considered in the rating of Item 60. Both existing and potential problems with scour should be reported so that a scour evaluation can be made by others. The scour evaluation procedure is reported on Item 113 in the revised "Bridge Recording and Coding Guide". If the bridge is determined to be scour critical, the rating of Item 60 should be evaluated to ensure that existing scour problems have been considered. The following items are recommended, based on my own experience, for consideration in inspecting the present condition of bridge foundations:

i) Evidence of movement of piers and abutments

- Rotational movement (check plumb line for straightness)
- Settlements (check lines of substructure and superstructure, bridge rail, etc. for discontinuities; check also for structural cracking or spalling)
- Check bridge seats for excessive movement

ii) Damage to scour countermeasures protecting the foundations (guide banks, sheet piling, sills, etc.)

iii) Changes in streambed evaluation at foundations (undermining of footings, exposure of piles)

iv) Changes in streambed cross section at the bridge, including location and depth of scour holes.

In order to evaluate the conditions of the foundations, the inspector should take cross sections of the stream, noting location and condition of streambanks. Careful measurements should be made of scour holes at piers and abutments, probing soft material in scour holes to determine the location of a firm bottom. If equipment or conditions do not permit measurement of the stream bottom, this condition should be noted for further action.

Assessing Scour Potential Bridges: The items provided in the Table "A" below are provided for bridge inspectors' consideration in assessing the adequacy of the bridge to resist scour. In making this assessment, inspectors need to understand and recognize the interrelationships between Item 60 (Substructure), Item 61 (Channel & Channel Protection), and Item 71 (Waterway Adequacy). As noted earlier, additional follow-up by others should be made utilizing Item 113 (Scour Critical Bridges) when the bridge inspection reveals a potential problem with scour.

Table A: Assessing the Scour Potential at Bridges

Upstream Conditions:

- Banks

  Stable: Natural vegetation, trees, bank stabilization measures such as paving, gabions, channel stabilization measures such as dikes and jetties.

  Unstable: Bank sloughing, undermining, evidence of lateral movement, damage to stream stabilization measures, etc.

- Main Channel
  - Clear and open with good approach flow conditions, or meandering or braided with main channel at an angle to the orientation of the bridge.
  - Existence of islands, bars, debris, cattle guards, fences that may affect flow.
  - Aggrading or degrading streambed.
  - Evidence of movement of channel with respect to bridge (make sketches, take pictures).

- Floodplain
  - Evidence of significant flow on floodplain.
  - Floodplain flow patterns – does flow overtop road and/or return to main channel?
  - Existence and hydraulic adequacy of relief bridges (if relief bridges are obstructed, they will affect flow patterns at the main channel bridge).
  - Extent of floodplain development and any obstruction to flows approaching the bridge and its approaches.

- Evidence of overtopping approach roads (debris, erosion of embankment slopes, damage to pavement, etc.).

- Debris
  - Extent of debris in upstream

- Other Features
  - Existence of upstream tributaries, bridges, dams, or other features, that may affect flow conditions at bridges.

Conditions at Bridge

- Substructure

- Superstructure
  - Evidence of overtopping by floodwater
  - Obstruction to flood flows
  - Design (Is superstructure vulnerable to collapse in the event of foundation movement, e.g., simple spans and nonredundant design for load transfer?)
- Channel Protection and Scour Countermeasures
  - Riprap (Is riprap adequately toed into the streambed or is it being undermined and washed away? Is riprap pier protection intact, or has it been removed and replaced by bed-load material? Can displaced riprap be seen in streambed below bridge?)
  - Guide banks (Spur Dikes) (are guide banks in-place? Have they been damaged by scour and erosion?)

- Stream and streambed (is main current impinging upon piers and abutments at an angle? Is there evidence of scour and erosion of streambed and banks, especially adjacent to piers and abutments? Has stream cross section changed since last measurement? If so, in what way?)

Waterway Area

Does waterway area appear small in relation to the stream and floodplain? Is there evidence of scour across a large portion of the streambed at the bridge?

Do bars, islands, vegetation, and debris constrict the flow and concentrate it in one section of the bridge or cause it to attack piers and abutments? Are approach roads regularly overtopped? If waterway opening is inadequate, does this increase the scour potential at bridge foundations?

Downstream Conditions

- Banks
  Stable:　Natural vegetation, trees, bank stabilization measures such as riprap, paving, channel stabilization measures such as dikes and jetties.

  Unstable: Bank sloughing, undermining, evidence of lateral movement, damage to stream stabilization measures, etc.

- Main Channel
  - Clear and open with good "getaway" conditions, or meandering or braided with bends, islands, bars, cattle guards, and fences that retard and obstruct flow.

- Aggrading or degrading streambed.
- Evidence of movement of channel with respect to the bridge.

- Floodplain
  - Clear and open so that contracted flow at bridge will return smoothly to floodplain, or restricted and blocked by dikes, development, trees, debris, or other obstructions.
  - Evidence of scour and erosion due to downstream turbulence.

- Other Features
  - Downstream dams or confluence with larger stream which may cause variable tailwater depths (This may create conditions for high velocity flow through bridge.).

Underwater Inspections: Perhaps the single most important aspect of inspecting the bridge for actual or potential damage from scour is the taking and plotting of measurements of stream bottom elevations in relation to the bridge foundations. Where conditions are such that the stream bottom cannot be accurately measured by rods, poles, sounding lines or other means, other arrangements need to be made to determine the condition of the foundations. Other approaches to determining the cross section of the streambed at the bridge include:

a) use of divers; and

b) use of electronic scour detection equipment

For the purpose of evaluating resistance to scour of the substructure under Item 60 of the Bridge Recording and Coding Guide, the questions remain essentially the same for foundations in deep water as for foundations in shallow water:

i) What does the stream cross section look like at the bridge?

ii) Have there been any changes as compared to previous cross section measurements? If so, does this indicate that (1) the stream is aggrading or degrading; or (2) local or contraction scour is occurring around piers and abutments?

iii) What are the shape and depths of scour holes

iv) Is the foundation footing (or the piling) exposed to the streamflow; and if so, what is the extent and probable consequence of this condition?

v) Has riprap around a pier been removed?

Notification Procedures: A bridge inspector's site evaluation of water at the bridge is an important part of a bridge inspection. A positive means of promptly communicating inspection findings to proper agency personnel must be established. Any condition that a bridge inspector considers to be of an emergency or potentially hazardous nature should be reported immediately. That information as well as other conditions which do not pose an immediate hazard, but still warrant further action, should be conveyed to the hydraulic or foundation engineers for review.

A report is, therefore, needed to communicate pertinent problem information to the hydraulic/geotechnical engineers.

ABOUT THE AUTHOR

H. John Parsaie earned his Ph.D. in Civil Engineering from the University of Berkley in Southfield Michigan. He has more than ten years of experience in the field of construction, quality control, consulting and services. Dr. Parsaie has extensive hands-on experience in construction materials testing and field inspection; construction management; project engineering; bridge design & inspection; as well as civil and structural design and project management. His experience also includes Forensics Engineering, site investigation, evaluation and analysis of structural and material failures, as well as monitoring unique construction procedures and methods. Dr. Parsaie has managed quality control, forensic engineering, site investigation and/or inspection and material testing programs for numerous major construction projects of various sizes and magnitudes, including projects outside the United States.

Dr. Parsaie has written, edited and published several books, papers and articles on engineering and management in various technical magazines including a four-part textbook series entitled "Construction Materials for Civil & Structural Engineering", which covers concrete, soils, steel, and plastics and a manual entitled "Training and Reference Manual for Special Inspectors".

Dr. Parsaie is serving as a consultant on the board of directors of many technical and professional societies, including the following:

➤ *International Conference of Building Officials (ICBO)*
➤ *American Welding Society (AWS)*
➤ *Society of American Military Engineers (SAME)*

➢ *International Association for Bridge & Structural Engineering (IABSE)*

➢ *National Society of Professional Engineers (NSPE)*

➢ *American Society for Testing and Materials, Member of the Committee C-7 (ASTM)*

➢ *National Academy of Forensic Engineers (NAFE)*

➢ *Washington Association of Building Officials, Member of the Sub-committee for Lateral Wood framing Special Inspection Program (WABO)*

APPENDIX

General Checklist for Concrete and Masonry Inspection

General Checklist for Structural Steel and Welding Inspection

General Checklist for Concrete and Masonry Inspection

❑ Check in with the general contractor, introduce yourself as special inspector.

❑ Check to see that the contractor maintains a copy of the approved plans on-site.

❑ Study the plans and specifications before starting inspection.

❑ Make sure that the contractor maintains appropriate permits on-site.

❑ Make sure that the contractor maintains copies of all approved mix designs.

❑ Check the mill certifications for steel, if applicable.

❑ Check the forms for adequacy of supports; check dimensions of the forms, etc.

❑ Check reinforcing steel for size, quantity, grades, clearances, etc.

❑ Check all embeds, straps, etc. for positioning, sizes, models, etc.

❑ Advise the contractor of any non-conforming work to be corrected prior to concrete placement. Please note that as a special inspector, you do not have the authority to stop the work. However, you may call the building official and report that the contractor's work is not in compliance with approved plans and you also may write a non-conforming report.

❑ Check delivery tickets of *all* concrete trucks for the correct mix design, etc.

❑ If the actual batch weights are not printed on the tickets, ask the driver to provide the batch weights and maximum allowable water that may be added to the mix.

❑ Visually inspect slump of all concrete batches delivered to the jobsite. However, make actual slump test on the first truck; additionally every fifth truck should be tested as well.

❑ Check the temperature of concrete; cast a minimum of set of three concrete specimens for each 150 cubic-yard of concrete placed, or as prescribed in the approved plans by the structural engineer.

❑ Add water to the mix, if needed, to the maximum allowable amount of water, and perform a slump test again, even though that you may have performed a slump test prior to adding water.

❑ When inspecting concrete masonry units (CMU), make sure that the masons are following the approved procedures that were discussed at the pre-construction meeting, including but not limited to the allowable lift heights, etc.

❑ Make sure that for concrete placements, the concrete is mechanically consolidated; masonry grouting is mechanically consolidated *and* reconsolidated.

❑ Write a report with findings and make sure that the report is signed by a representative of the general contractor, especially when writing a non-conforming report.

General Checklist for Structural Steel and Welding Inspection

❑ Check in with the general contractor, introduce yourself as special inspector.

❑ Check to see that the contractor maintains a copy of the approved plans on-site.

❑ Study the plans and specifications before starting inspection.

❑ Make sure that the contractor maintains appropriate permits on-site.

❑ Check the mill certifications for steel, if applicable.

❑ Make sure that the contractor is installing the appropriate bolts, dowels, etc.

❑ Perform torque testing on bolts and nuts as appropriate.

❑ Check the qualifications of the welders on the jobsite and/or fabrication shop.

❑ Make sure that the contractor has a welding procedure in-place that conforms to the American Welding Society (AWS) standards.

❑ If necessary and allowed by code and building official, qualify welders for the work that is being performed using AWS procedures.

❑ Write a report with findings and make sure that the report is signed by a representative of the general contractor, especially when writing a non-conforming report.

REFERENCES

Building Codes:

- 1997 Edition of Uniform Building Code (UBC)
  - Volume 1   Chapter 1–Administration
  - Volume 2   Chapter 16–Structural Design Requirements (Division I, IV, V,); Chapter 17–Structural Tests and Inspections (Section 1701, 1702, 1703, 1704); Chapter 19–Concrete (Division I–VIII); Chapter 21–Masonry (All Sections); Chapter 22–Steel (All Divisions); Chapter 23–Wood (Division I–VIII)
  - Volume 3   UBC Standard 7-6–Thickness, Density Determination and Cohesion/Adhesion for Spray-applied Fire-resistive Materials

- 2000 Edition of International Building Code (IBC)
  - Chapter 1–Administration; Chapter 16–Structural Design Requirements; Chapter 17–Structural Tests and Inspections; Chapter 19–Concrete; Chapter 21–Masonry; Chapter 22–Steel; Chapter 23–Wood

American Welding Society Publications:

- AWS–D 1.1:2000   Structural Welding Code–Steel
- AWS–B 2.1:1998   Specification for Welding Procedure and Performance Qualification
- AWS–FMC:2000   Filler Metal Comparison Charts

- AWS–D 1.4:1998   Structural Welding Code–Reinforcing Steel
- AWS–D 1.3:1998   Structural Welding Code–Sheet Steel
- AWS–QC-G   Guide to AWS Welding Inspector Qualification and Certification
- AWS–D 10.13-95   Recommended Practices for the Brazing of Copper Pipe & Tubing Medical Gas Systems
- AWS–B 2.2-91   Standard for Brazing Procedure and Performance Qualification
- AWS–C 3.5:1999   Specification for Induction Brazing
- AWS–A 5.31-92   Specification for Fluxes for Brazing & Braze Welding

Federal Emergency Management Agency

- FEMA 267–Interim Guidelines: Evaluation, Repair, Modification & Design of Welded Steel Moment Frame Structure
- FEMA 267A–Interim Guidelines–Advisory No. 1–Supplement to FEMA 267

Structural Engineers Association of California

- SAC 95-01–Steel Moment Frame Connection (Advisory No.3–Supplement to FEMA 267)

Concrete Reinforcing Steel Institute

- MSP-2-98   Manual of Standard Practice

American Concrete Institute

- ACI–Manual of Concrete Practice, Part 1–5
- ACI Manual of Concrete Inspection (SP-2-99)
- ACI 318-95–Building Code Requirements for Structural Concrete & Commentary (318R-95)
- ACI 301-96–Structural Concrete Specifications

Masonry Institute of America

- Reinforced Grouted Brick Masonry
- Reinforced Masonry Engineering Handbook
- Masonry Codes & Specifications

Precast/Prestressed Concrete Institute

- Manual for Quality Control
- PCI–Design Handbook

Post –Tensioning Institute

- PTI–Specification for Unbounded Single Strand Tendons
- Post-Tensioning Manual

American Society for Testing and Materials

- ASTM C29–Test method for unit weight and voids in aggregate
- ASTM C31–Practice of making and curing concrete test specimens in the field

- ASTM C39–Test method for compressive strength of cylindrical concrete specimens
- ASTM C40–Test method for organic impurities in fine aggregates for concrete
- ASTM C42–Test method for obtaining and testing drilled cores and sawed beams of concrete
- ASTM C88–Test method for soundness of aggregates by use of sodium sulfate or magnesium sulfate
- ASTM C117–Test method for materials finer than No. 200 sieve in mineral aggregates by washing
- ASTM C127–Test method for specific gravity and absorption of coarse aggregate
- ASTM C128–Test method for specific gravity and absorption of fine aggregate
- ASTM C136–Test method for sieve analysis of fine and coarse aggregates
- ASTM C138–Test method for unit weight, yield, and air content of concrete
- ASTM C143–Test method for slump of hydraulic cement concrete
- ASTM C172–Practice for sampling freshly mixed concrete
- ASTM C173–Test method for air content of freshly mixed concrete by the volumetric method
- ASTM C192–Practice for making and curing test specimens in the laboratory
- ASTM C231–Test method for air content of freshly mixed concrete by the pressure method
- ASTM C293–Test method for flexural strength of concrete

- ASTM C490–Practice for use of apparatus for the determination of length of hardened cement paste, mortar, and concrete
- ASTM C617–Practice of capping cylindrical concrete specimens
- ASTM C803–Test method for penetration resistance of hardened concrete
- ASTM C805–Test method for rebound number of hardened concrete
- ASTM C1064–Test method for temperature of freshly mixed Portland cement concrete
- ASTM C1140–Practice for preparing and testing specimens from shotcrete test panels
- ASTM C1231–Practice for use of unbounded caps in determination of compressive strength of hardened concrete cylinders
- ASTM E1155–Test method for determining floor flatness and levelness numbers
- ASTM E1703–Test method for measuring rut-depth of pavement surfaces using straightedge
- ASTM C67–Test method of sampling and testing brick and structural clay tile
- ASTM C140–Test method of sampling and testing concrete masonry units
- ASTM C109–Test method for compressive strength of hydraulic cement mortars
- ASTM C1019–Test method for sampling and testing grout
- ASTM C1314–Test method for construction and testing masonry prisms used to determine compliance with specified compressive strength of masonry
- ASTM C29–Test method for unit weight and voids in aggregate

- ASTM D75–Practice of sampling aggregates
- ASTM D546–Test method for sieve analysis of mineral filler for road and paving materials
- ASTM D979–Practice for sampling bituminous paving mixtures
- ASTM D1188–Test method for bulk specific gravity and density of compacted bituminous mixtures using paraffin-coated specimens
- ASTM D1559–Test method for resistance to plastic flow of bituminous mixtures using marshall apparatus
- ASTM D2041–Test method for theoretical maximum specific gravity and density of bituminous paving mixtures
- ASTM D2419–Test method for sand equivalent value of soil and fine aggregate
- ASTM D422–Test method for particle-size analysis of soils
- ASTM D698–Test method for laboratory compaction characteristics of soil using standard effort
- ASTM D854–Test method for specific gravity of soils
- ASTM D1140–Test method for amount of material in soils finer than No. 200 sieve
- ASTM D1556–Test method for density and unit weight of soil in-place by the sand cone method
- ASTM D1557–Test method for laboratory compaction characteristics of soil using modified effort
- ASTM D1883–Test method for CBR (California Bearing Ratio) of laboratory compacted soils
- ASTM D2166–Test method for unconfined compressive strength of cohesive soil
- ASTM D2216–Test method for laboratory determination of water (moisture) content of soil and rock

- ASTM D2435–Test method for one-dimensional consolidation properties of soils
- ASTM D2922–Test methods for density of soil and soil-aggregate in-place by nuclear methods
- ASTM D2937–Test method for density of soil in-place by the drive-cylinder method
- ASTM D2938–Test method for unconfined compressive strength of intact rock core specimens
- ASTM D3017–Test method for water content of soil and rock in-place by nuclear methods
- ASTM D3080–Test method for direct shear test of soils under con-solidated drained conditions
- ASTM D4318–Test method for liquid limit, plastic limit, and plas-ticity index of soils
- ASTM E114–Practice for ultrasonic pulse-echo straight-beam examination by the contact method
- ASTM E164–Practice for ultrasonic contact examination of weld-ments
- ASTM E165–Test method for liquid penetrant examination

Department of Transportation

Washington State Department of Transportation–Bridge Inspection Manual

U.S. Department of Transportation–Scour Evaluations